T0253162

SIMPLE WORKSHOP DEVICES

Tubal Cain

Special Interest Model Books

Special Interest Model Books Ltd.
P.O.Box 327
Poole
Dorset
BH15 2RG
England

First published by Argus Books Ltd. 1983
Second edition published by Nexus Special Interests Ltd. 1998
This edition published by Special Interest Model Books Ltd. 2005
Reprinted 2008, 2011, 2015

© Special Interest Model Books Ltd. 2005

The right of Tubal Cain to be identified as the Author of this work has been
asserted by him in accordance with the Copyright, Designs and Patents Rights
Act of 1988.

All rights reserved. No part of this book may be reproduced in any form by
print, photography, microfilm or any other means without written permission
from the publisher.

Cover photo by Mike Chrisp

ISBN 978-1-85486-150-4

www.specialinterstmodelbooks.co.uk

Printed and bound in Malta by Melita Press

Contents

Introduction

Reprinting this title to bring it into the highly successful *Workshop Practice Series* has given me the opportunity both to introduce a number of new devices and to revise some of the original matter but one important change – metric equivalents to imperial dimensions – has not been made nor have Morse number drill sizes been corrected to millimetres. To have done this without cluttering up the drawings would have required that almost all of them be retraced. I have, therefore, assumed that readers are quite capable of making the conversions themselves where necessary.

I should, perhaps, explain the use of the word 'simple' in the title and the philosophy that lies behind it. I have the greatest respect for those who design and build sophisticated workshop equipment and do, in fact, take advantage of their work – it would be a poor workshop which had no tool-and-cutter grinder! But my interest lies in the designing and making of models of engines, real or imagined, or things for them to drive.

So, if any 'manufacturing device' is needed I always carry out a study of its cost effectiveness (cost in my case being mainly the time involved to make it) and unless any such device will either save a lot more time or will facilitate accurate working – preferably both – I will tend to disdain it. I try to look for solutions to problems which are both simple in conception and speedy in construction. Most of the ideas shown in this book can be made in a few hours (sometimes even minutes) and few require bought-in material.

In the introduction to the first edition I mentioned (at the suggestion of friends!) a sort of pecking order for the various devices, the implication being that these should be made first, but this does present many difficulties. My most used device – a little piece of wood or a length of string to hold bits together – is not shown! However, the following may help.

Over the years since the first edition was published, the **centre-height gauge** has been the most used by far, both for tool-setting and for marking out. As frequently mentioned in my other books, I now rough screwcut before using the tailstock die-holder whenever I can. The **depth stop** is invaluable and has also come in useful for milling flutes and cutting small gears. **Tool-posts** – both the **Gibraltar** and that for

1

The author at work!

hand turning – I would most certainly replace if I lost them. The **overhead drive** is not often used but how could I manage without it? I doubt if a week passes by without recourse either to a **cross-drilling jig** or the **filing buttons**. In short, the very fact that I have mentioned a device in this book implies that I have need of it!

Finally, to repeat myself, there is no single, unique solution to any engineering problem – that is what makes the profession so much fun. Do not be afraid to modify either design, dimensions, materials or construction methods to suit your own circumstances. Some of the devices described were made before I had even a vertical slide, let alone a milling machine!

Tubal Cain
Westmorland, 1998

The 'heavy machining' end of the workshop.

Workbench with storage under and the 'light work' area on the right.

The lathe – the 'queen' of machine tools.

The brazing bench, fitted with a turntable, which is sited outside the main workshop area (Photo Mike Chrisp).

Getting Hold of the Job

A proper hold of the workpiece is fundamental to all manufacture, whether of models or in full-scale production. But we have a rather special problem as many components are either too small or too thin to grip firmly. We cannot afford either the time or the money to invest in the sophisticated workholding devices used in industry.

In this short section I have not covered the obvious accessories like chucks and bench vices, nor jigs and fixtures which come later. The former are normal workshop equipment and the latter are usually special for each job that crops up. However, I hope that the following pages may save a few workpieces, and perhaps also fingers, from damage. More important, they may suggest to you other ways of tackling those awkward jobs; if so, please don't keep quiet about them – make sketches, take a photograph and send a note to the editor of *Model Engineer* (also published by Nexus Special Interests).

The thinpiece vice
The holding of thin material in the vice has always presented problems. The unit described here will overcome most of them for material down to about $\frac{1}{32}$in. thick. It was originally devised as an apprentice training exercise by the supervisor of the training school of a large firm, and has given me excellent service for many years. The design has been simplified for the model engineer's use, so that the drawings and the photographs do not quite correspond (e.g. items 4 and 5 are milled from a single piece of material in the original). See Fig 1.1.

Make the jaws, items 4 and 7 (Fig. 1.2) first, as these will be required as gauges for the top plate and body. The ideal material is $\frac{1}{2}$ in. square ground gauge stock as this needs no preparation. Failing this, take an $8\frac{1}{2}$ in. piece of $\frac{1}{2}$ in. material and carefully file parallel and square. Keep the width to a uniform thickness – plus or minus 0.001 in. The depth is not so critical. File one end square, cut off a piece $4\frac{1}{16}$ in. long and repeat the process. (The squared ends will ultimately be the *top* of the jaws.) Coat with copper sulphate or marking blue and mark out all holes and the slots in the lower ends, taking all dimensions from the squared ends.

Tackle the slots before the rest, so that if this job goes awry as little work

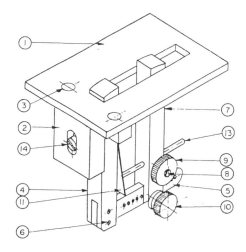

Fig. 1.1 Thinpiece vice.

as possible is wasted. As an apprentice exercise these are made using hacksaw and file, which is probably quicker than attacking them with a $\frac{3}{16}$ in. slot drill, but if a $2\frac{1}{2}$ in. $\times$ $\frac{3}{16}$ in. slitting saw is available this will make short work of the cut. Run at about 60 rpm and feed the work in steadily, using plenty of cutting oil. Finish off with a 4 in. warding file if need be, making the front jaw a tight fit on a $\frac{3}{16}$ in. thick gauge (e.g. the material for part 5) and the rear jaw a sliding fit. Remove burrs.

Drill and tap the holes in the front jaw, but leave the two No. 31 holes at the bottom for the present. The exact shape of the 'oval' hole in the rear jaw is unimportant, as part 8 can be made to fit. The quickest way is to drill two $\frac{3}{16}$ in. holes at $\frac{5}{16}$ in. centres and file out the remainder, but professionals will doubtless use patience and a slot drill! Do not drill the $\frac{5}{32}$ in. cross hole yet. Drill No. 37 as shown, open out one side to $\frac{1}{8}$ in. and tap the other side 5BA using the $\frac{1}{8}$ in. hole as a guide.

Now hold the jaws side by side in the vice – making sure they are the right way round – and file the previously squared ends to the 2 in. radius shown. The bottom end of the rear jaw may now be radiused, but leave the front jaw for the present.

The top plate, item 1, is best made from ground flat stock, but mild steel plate will serve if it is truly flat. File to shape and mark out the slot and the two rivet holes, taking care that the $\frac{5}{32}$ in. dimension is reasonably accurate. Drill a series of $\frac{3}{16}$ in. holes (1,000 rpm if ground stock, 2,000 rpm if BMS for HSS drills) well inside the lines of the slot and file until the jaws are a nice sliding fit. See that the ends of the slot are square. Start the rivet holes with a Slocombe drill, but do not drill through.

The body, item 2, calls for some energetic work with hacksaw and file to bring to the T shape. It is important that the top face be flat, and square to the front face with the $\frac{1}{2}$ in. groove in it. Mark out the front face for this groove, set up on the vertical slide with a piece of packing behind, so that the groove may be milled using the saddle cross slide, and make sure (a) that the $2\frac{3}{8}$ in. wide top face is square to the lathe bed and (b) that the front face is square across the bed. Using a $\frac{3}{8}$ in. slot drill at about 650 rpm take a full depth cut across the centre of the slot and then work carefully towards the two lines with light cuts until the front jaw is a nice sliding fit in the groove.

Now adjust the vertical slide until the centre of the groove is at exact centre height, and with a Slocombe drill in the chuck start a hole $\frac{9}{16}$ in. from the top face. Follow this with a $\frac{3}{16}$ in. drill to make a hole right through (hence the need for packing behind). Chuck a $\frac{3}{16}$ in. slot drill and cut the $\frac{3}{16}$ in. slot as shown, slightly

6

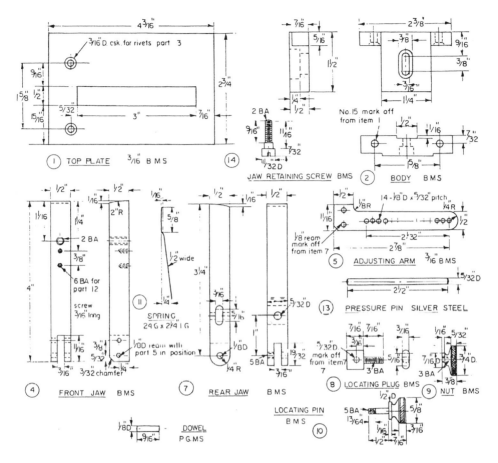

Fig. 1.2 Details of thinpiece vice.

deeper than the drawing requires. Remove the work, and transfer the centre of the slot to the other side of the body. Set up on the vertical slide, squaring up as before; line up with the $\frac{3}{16}$ in. hole. Drill a $\frac{3}{8}$ in. hole on the centre until it just breaks into the $\frac{3}{16}$ in. slot and follow with a $\frac{3}{8}$ in. slot drill to machine the slotted counterbore (if your drills have a tendency to run oversize holes, use a No. 14 and letter U instead of $\frac{3}{16}$ in. and $\frac{3}{8}$ in.).

Use plenty of cutting oil during the milling operations, and keep a steady feed allowing no rubbing without cutting.

The adjusting arm, part 5, is a simple filing and drilling job, best made of ground flat stock, but BMS flat will do. The row of holes can be drilled very accurately by mounting the arm (with packing behind) on the vertical slide. With a Slocombe drill in the chuck,

7

deeply drill the first hole; advance the cross slide 156 thou, drill the next, and so on. The holes may then be finished to $\frac{1}{8}$ in. in the drilling machine, but the pitch and alignment will be really true. Lightly countersink both sides. Do not drill the two No. 31 holes yet.

The two rivets, item 3, are turned from soft mild steel or from a longer $\frac{5}{16}$ in. iron rivet. The dowels, item 6, are parted off from ground BMS rod, **not** silver steel, and the pressure-pin, item 13, from silver steel. The locking plug, item 8, is made by chucking a piece of $\frac{3}{16}$ in. × $\frac{1}{2}$ in. steel in the 4-jaw, turning the screwed part 0.160/0.161 in. dia. and screwing 3 BA with the tailstock die-holder. The rectangular plug is then filed a nice fit into the hole in the rear jaw. Assemble into the jaw with the faces 'A' flush, and drill $\frac{5}{16}$ in. right through. (Purists may prefer to drill No. 24 and ream, but this is not necessary.) Lightly countersink the entrances to the holes.

The nut and pin, items 9 and 10, are simple turning jobs, the exact shape of the heads being left to individual taste but knurl *before* parting off in both cases. The screw, No. 14, may be a standard Allen socket-head, if desired.

It should be possible to put the kink in the spring, item 11, without softening it, if held in the smooth jaws of the vice. If the holes are to be drilled, run no faster than 360 rpm, hone the drill point so that it is really sharp; rest the work-piece on a piece of steel packing and on no account hold the spring in the hand while drilling. It may be simpler to punch the holes with a sharp flat punch on a lead anvil.

Attach the front jaw 4 to the body 2 with a piece of paper (or .002 in. shim) in the bottom of the groove, using the screw 2. Clamp the top 1 to the body,

ensuring that the jaw is hard up against the end of the slot. Drill and ream the two $\frac{3}{16}$ in. fixing holes. Dismantle and countersink the holes as shown and remove burrs. Insert the rivets 3, rivet up and carefully file flush. File the end of the slot in the top plate so that it is flush with the groove. It may be necessary to ease the slot a trifle to ensure that the jaw can slide up and down.

Insert the arm 5 into the slot in the jaw 4, first smearing some *Easyflo* paste flux on the mating parts. Check that it is square to the jaw. Drill and ream the upper hole; press in one of the dowels 6 and rivet lightly. Check for squareness, drill and ream the second hole and fit the pin as before. Heat to dull red, apply *Easyflo* silver solder, allow to cool to black, and quench in cold water.

File the lower end of the jaw to the profile of the arm, and smooth off the projecting pins. Polish off with fine emery. Attach the spring. Push the plug 8 into its place in the jaw 7, insert the pin 13 and lightly tighten the nut 9. Assemble the jaw to the arm 5 with the pin 10. The whole jaw assembly is attached to the body by the screws 14.

In service, the unit is held in the bench vice (with fibre grips in place) with the jaw 4 to the front, Fig 1.3. The projection of the jaws is adjusted by means of the screw 14 to suit the thickness of the job to be held. The rear jaw position is adjusted on the arm to suit the width of the workpiece, and the pressure pin positioned so that the spring opens the jaws to follow those of the bench vice.

My own vice has been in fairly constant use for 25 years. It *would* pay to harden the ends of the moving jaws. This can be done by casehardening; heat up to red and cover the top half-

Fig. 1.3 An underside view of the vice.

inch with *Kasenit* compound. Reheat to red again (see the instructions on the tin) and quench in water. There is no need to get the whole of the arm hot, of course. This operation should be done *before* uniting parts 4 and 5. After cleaning up and polishing this joining can be done but, while brazing, immerse the lower ends (that is, the hard tops) in a water bath. This will prevent drawing the hardness of the nose of the jaws.

Although the top plate has got a little scarred with use I doubt if it is worth hardening this, although if gauge plate is used it can be done by heating to about 800°C for say ten minutes (cherry red) and then quenching in oil, vertically. Temper to pale straw. The risk is

Fig. 1.4 In service – though the workpiece is rather thick!

9

that the plate may distort, and there is the added point that a hard plate will damage the files. My own is soft gauge plate and the way things are going I may have to make a new top in ten or a dozen years' time; cheaper than new files!

Circular work in the drilling vice
All the books tell you that work should be clamped to the table or held in a vice while drilling, and this applies especially when working in brass or gunmetal. The material drags at the drill point and unless you have the old-fashioned straight flute drills (or the modern, and expensive, slow helix type) a 'snatch' when breaking through is inevitable. (You ought, of course, to take off the rake at the drill point when drilling brass and then resharpen for normal work, which means that after 12 months or so all your drills will be tiny stubs of HSS!.) So, drilling vice or clamp it must be. The snag comes when holding small round objects; stuffing box glands, for example. Even if your vice is furnished with vees on the jaw faces there is the risk of marking the carefully turned stem of the gland and if you don't grip

hard enough the drill will pull the job out of the vice and do more damage still; to you or to the job!

Fig. 1.5 shows how I got over this problem. The block is made from a piece of close-grained hardwood. Do not use oak – beech should do, and yew better still. But I use lignum vitae, which can be obtained from worn-out bowls woods or old foundry rammers. A good alternative is boxwood if you can get any. A series of holes, corresponding to the most common diameters you find in your work, are drilled through, the block having first been squared all over, of course. A small hole, I suggest $\frac{1}{8}$in., is drilled as shown and the block then sawn through as far as this hole. It is held between the jaws of the vice and will grip the workpiece well if put in the right hole. There is, of course, no reason why it should not be cut into two separate pieces (it may break in two eventually) but I find it convenient to have it as shown in Fig. 1.5.

Fig. 1.6 shows a variation for holding such things as engine cylinder covers etc. I make mine to fit the OD of the flange as getting a better grip than on the spigot only. This one is bored in the

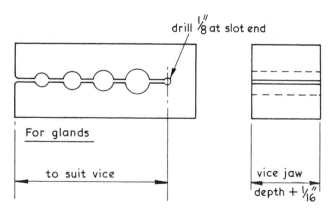

drill ⅛" at slot end

For glands

to suit vice

vice jaw
depth + ¹/₁₆"

Fig. 1.5 Holder for glands and similar small diameter work.

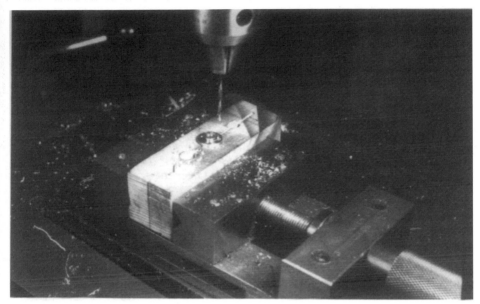

Fig. 1.5a Drilling a gland flange.

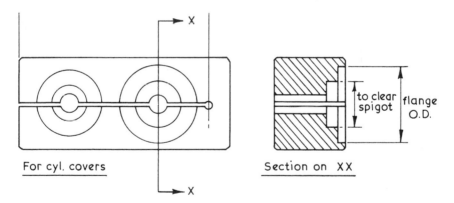

For cyl. covers

to clear spigot | flange O.D.

Section on XX

Fig. 1.6 Holder for discs and similar work.

lathe to suit the job in hand, and I make the block quite wide so that small cavities can be opened up if necessary. Naturally, either type gets riddled with holes after a while. That for the glands and small objects can be used both sides of course, but even so will need replacing every few years. Why not make them of metal? No reason at all, except that wood is cheaper and quicker

11

to work. In passing, lignum is a real engineer's wood. As you may know, it is used for stern-tube bearings on steamships and the old Alvis *Firefly* had lignum bearings in all the steering joints. So, you might care to consider using it as a chuck fixture when holding cylinder covers in the lathe, instead of making up a steel or brass split collet. I do and it is perfectly satisfactory.

Thinning washers (Turner's cement)

'Making big ones into little ones' was once the standard punishment for those sentenced to hard labour, and this work is still a penance to the model engineer, especially when it comes to washers or similar round objects which must be thinned. The problem is holding the things – or indeed any thin object of irregular shape. The thinpiece vice (page 5) doesn't help and to alter it to hold washers still leaves the other awkward shapes which turn up every now and then. I suppose that everyone knows the dodge of pressing the article into a piece of softwood using the vice jaws, so that it makes itself a recess in which it can be held while filing. It is indeed very effective, and I have lots of bits of wood lying around (kept just in case another job of the same shape crops up) that testify to this! Opinions differ as to the best wood to use, but I find offcuts of yellow pine the best, with the work pressed into the side grain rather than the end grain as suggested in some books.

However, the method has its limitations, the chief being that it is difficult to get an even thickness and almost impossible to achieve any degree of precision. This is where the substance known as **Turner's cement** comes in. There are many recipes ranging from neat shellac to a horrible mixture of resin and pumice powder. The easiest to make and use, and almost the strongest, is three parts of common resin mixed with one part beeswax. The wax is melted (with due precautions against it catching fire) and the resin, preferable in very small pieces, dropped in. The mixture is then heated a little longer and well stirred until it is seen that the two ingredients are both melted and well mixed. Cast it into sticks – I simply pour it into the vee of a piece of (clean) angle iron and then break up the length into convenient sizes.

It can be used in many ways. For woodwork the stick is held against the rotating faceplate until friction melts it on to the face. The workpiece is then similarly held against the cement coat until it again melts and then, when increasing resistance is felt, pressure is reduced until the work 'catches'. Then let go, and you will find the work is securely held – it is as easy as that! For metal work this is rather hazardous, and I apply heat to the holding piece (faceplate or whatever) until the cement melts on to it in an even film. Then apply the warmed workpiece and heat again if necessary to get an even thickness. You must then let it cool, or hold it under the cold tap, until the stuff has set. I have a spare lathe backplate I use, but equally often I simply chuck a piece of scrap, face it true and use that. Fig. 1.7 shows half a dozen washers being so treated. These had to be thinned down to 0.020 in. thick, and cuts of about 5 thou were taken with no difficulty. All were equal in thickness within much less than half a thou. After machining they were removed simply by tapping them with a plastic mallet, although they could have been melted off if need be.

I have used this method in place of the 'block of wood', holding a flat chunk

Fig. 1.7 Six washers attached to a carrier plate with Turner's cement.

of cast iron in the vice, and on the lathe have faced down rings of up to 6 in. diameter. The grip is good (about 200 lbf/sq.in. of contact area on test) but where there may be an interrupted cut care must be taken, as the cement won't stand heavy shock loads. You will notice that in the photo the washers are all touching each other, in a ring, so that all help to take the shock as the tool passes over the holes. The method is, of course, a variant on the use of solder to hold things, but is less messy, doesn't need so high a temperature, and is much easier to clean off afterwards.

The cement can be used to hold pieces while milling, but both cut and feed rate must be small. I have used it on the drill, and no problems arose except that occasionally the work came adrift from the backing piece on breaking through. The one thing that must be watched – obvious if you weigh it up – is that you must not get the workpiece hot during machining! Coolant will have no effect on the cement.

A soldering and brazing clamp

Here is a little gadget you can make in an hour or so which will save you endless time and frustration for the next twenty years! The photograph on page 14 shows the device in use, after brazing a butt joint between a 16 gauge brass wire and a ring $\frac{1}{8}$ in. wide × 28 thou thick – a job which normally means endless fiddling and even then often comes out all wrong.

I lay no claim to originality for the design. The incomplete 'bits' have been sculling around my workshop for many years and only a recent gap in the production programme made me think of making up the missing parts 'in case it came in useful', which it did within a day or so, and I now regret that I didn't put it in order years ago.

The drawing, Fig. 1.9, shows the details. The clamping arms are the most awkward part, as they must be springy. The one remaining original was of a very hard brass, and to get anywhere near the elasticity I had to work-harden a piece of half-hard brass strip by hammering it well. Possibly drawn brass might be hard enough as an alternative.

Cut out a strip about 9 in. long, say 16 gauge, and planish it until you have reduced the thickness by at least three thou. If you haven't a proper planishing hammer (which has a slightly convex face) you will have to use care and the flat of your light riveting hammer. Keep the work flat on a suitable anvil and ensure that the face of the hammer strikes flat as well. Use a multitude of medium-light blows – don't give it a belting – and take care to get the impacts spread over the whole of the surface. Turn the job over periodically so that each face gets even treatment.

If the strip starts to bend sideways

13

Fig. 1.8 Brazing a butt joint between a wire and a narrow ring.

this means you are hitting one side of the strip more the other; give the next set of blows on the concave side to correct it.

Cut off into 2 in. lengths (you need four) and mark out for the holes, noting that there are four holes in two arms and three in the other. Drill the $\frac{3}{32}$ in. hole in the end, clamp all four arms together with a pin in this hole, and drill through 6 BA tapping size the holes common to all arms; then open out the clearing holes and the additional tapped holes in the 'top' arms, finally tapping as shown. The countersink on the end hole is to embrace the ball on the clamp head. I used a $\frac{5}{16}$ in. drill point and this seems to work adequately.

For the thinned part in the middle, make a groove with a 3-cornered file on the centre, $1\frac{3}{32}$ in. from the end. Enlarge this with a rat-tail file, and then proceed with about a 10 in. half-round file, which will give about the right degree of curvature.

The inner face of the last $\frac{3}{8}$ in. at the nose of the clamp is serrated – done either by making a pattern with your centre punch, or by criss-crossing with a sharp chisel and light hammer blows.

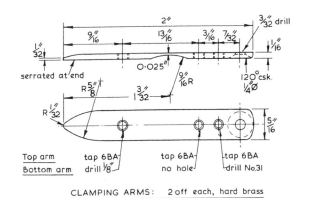

serrated at end
O·025″
Top arm
Bottom arm

tap 6BA
drill ⅛″

tap 6BA
no hole

tap 6BA
drill No.31

CLAMPING ARMS: 2 off each, hard brass

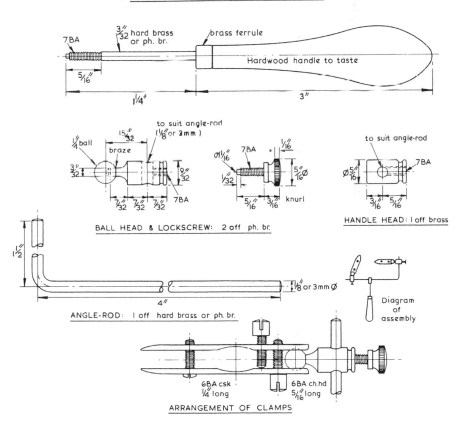

7BA
3/32″ hard brass or ph. br.
brass ferrule
Hardwood handle to taste
5/16″
1¼″
3″

to suit angle-rod (⅛″ or 3mm)
¼″ ball
15/32″
braze
3/32″
9/32″
7/32″ 7/32″ 7/32″
7BA

Ø1/16″
7BA
1/16″
1/32″
5/16″Ø
5/16″ 3/16″ knurl

to suit angle-rod
7BA
Ø5/16″
3/16″ 5/32″

BALL HEAD & LOCKSCREW: 2 off ph. br.

HANDLE HEAD: 1 off brass

1½″
4″
⅛″ or 3mm Ø

ANGLE-ROD: 1 off hard brass or ph. br.

Diagram of assembly

6BA csk
¼″ long
6BA ch.hd
5/16″ long

ARRANGEMENT OF CLAMPS

Fig. 1.9 Arrangement and details of the clamp set.

15

Do this after forming the profile in plan, but before tapering the thickness. The 'heel' of the clamp (the other end, that is) must be rounded a bit to clear the ball head when at an angle. Finally, finish with fine emery and polish.

The ball head on the original (there was only one remaining) was turned from the solid, and you can do the same if you like. As a preliminary measure I cheated a bit, and used an old terminal from an ancient wireless set, but it didn't fit the rod very well. So I made a new one and drilled a $\frac{1}{4}$ in. bronze ball to about half diameter, finally brazing this to the complete body of the part. The only point about this component is that the cross-drilling must be accurate – use one of the many drilling jigs described later in the book – and the hole marked 'to suit rod' must be a good sliding fit.

The lockscrew is a fancy knurled knob, but it would work just as well with a plain cheese-head screw. Don't omit the little pip on the end which prevents the thread from being burred over in use. I have specified phosphor bronze, as the thread will get a lot of use.

The handle head is a repeat of the ball head but without the ball, and with a longer threaded part. There is no reason at all why you shouldn't make the ball head $\frac{5}{16}$ in. instead of $\frac{7}{32}$ in. here – mine had to be as shown to match the existing one. The handle itself screws into the head, and locks on to the angle rod (see the little line sketch of the assembly) and can be made to suit your taste or your hands. It could, of course, be simply a piece of brass hex stock, to hold in the vice, but experience shows that it's easier to use if held in the hand against a fixed source of heat rather than applying the flame to the fixed job.

It is much easier to get the flame where you want it if you can move the work as you please.

To fit the rod into the handle – find a nice piece of rosewood or ebony – drill a hole in the handle itself slightly larger than the brass rod; introduce some *Araldite* and then push in the rod, and allow to cure. Don't forget the little brass ferrule.

I have left the angle rod until last. Mine is of very hard brass, and to avoid springing in use you need something similar. A length of *Sifbronze* or whatever will not be good enough. My suggestion for this part is that you make the rod up from $\frac{1}{8}$ in. or 3mm silver steel (drill the heads to suit) and after bending it, harden and temper to blue. I suggest, too, that you treat yourself to two rods, one bent as shown, and one straight, about $4\frac{1}{2}$ in. long.

The arrangement of the screws in the clamps is shown in the sketch. The 6 BA countersunk screw is *not* in a countersunk hole, but the clearance is rather greater than normal. (I have shown $\frac{1}{8}$ in. drill, but it could be a shade larger.) The other two screws are ordinary cheese-head brass with the heads reduced a trifle in diameter for appearance sake. All three screws must be adjusted when making any marked change in the width of grip of the clamp, but for small alterations it is only necessary to adjust the countersunk one. The clamps can be turned every which way on the balls, and the grip on these balls should be enough to 'stay put' but still be movable by hand pressure. You may need a little experiment to get the tensions right, so that the work is held with just the right pressure.

The points of the clamps *will* need some adjustment. Grip a piece of say

$\frac{1}{16}$ in. rod, and make sure that the clamp holds it across the full face; if it doesn't, one or both clamp arms are twisted in their length, and this should be rectified. I found, too, that it gave a better grip if the nose of the clamps was opened out a trifle, so that when gripping a piece of $\frac{3}{32}$ in. sheet the faces bedded flat.

The device is, of course, for light articles only – it's not meant for use in 5 in. gauge boilersmithing! It is, in fact, a silversmith's clamp, and used by allied trades such as watchmakers and jewellers. I have successfully brazed together two $\frac{1}{32}$ in. wires with it, and one unexpected use has been the joining together of short lengths of silver solder. You could, of course, make a larger version and arrange it on a bench stand, similar to those used in the old days by plumbers when jointing lead pipes. But I find it is the *little* jobs that cause the trouble; heavier items usually stay put when brazing under their own weight, or can be bound together with iron wire.

Finally, why not make it all of steel? No reason at all except that mine was originally of brass, and it does look nice when it's clean!

Packing for the machine vice

It is unfortunate, but when making (or buying) 'parallels' we tend to go for round numbers of fractions e.g. $\frac{3}{4} \times \frac{3}{8}$ in. This means that if we use such packing on edge to hold $\frac{3}{8}$ in. stock the vice tends to grip the packing rather than the workpiece. In this case $\frac{23}{32} \times \frac{11}{32}$ would be more convenient (precision-ground parallels for use in marking out are a different matter). You should, of course, use slips of paper between both jaws and work and between packings, to improve the grip and to take care of minute irregularities, then give the workpiece a thump with a mallet to bed it down properly.

Do not, however, be tempted to thump the vice handle as well 'just to make sure'. The mechanical advantage of most milling vices is of the order of 200/1, and even a moderate blow (which is a shock load) on the handle could impose an extra force of ten tons or so on the workpiece. If the vice seems to need such treatment to grip properly then it is overdue for an overhaul.

Jigs and Fixtures

These are really holding devices and could, perhaps, have been included in the previous section of the book. However they do have one significant feature – as a rule, they are all made for a *particular* workpiece and have not the universal application of a vice or clamping device. The distinction is somewhat blurred, but in general a jig sits on, or is clamped to, the work and is used to locate holes, cutters, or perhaps mating parts during assembly. A fixture normally holds the workpiece in a particular position or attitude so that cutting tools can be applied in a convenient manner. Some fixtures are also jigs, and vice versa, and the word jig is often applied quite indiscriminately to either.

It would be impossible to describe a jig for every circumstance, not even to describe those I use, for I have scores, if not hundreds, rigged up over the years. But the examples I give will, I hope, suggest ways of designing others to suit your own particular needs. Part of the fun of model engineering is finding out new ways of doing things. The first few pages of the section are concerned with the principles, and especially with the accurate location of holes. The remainder are more or less special-purpose jigs, but which have very general application. I have, at the end, included a single-purpose fixture, very simple indeed, which saved me a lot of money, a matter not to be disdained these days!

Why use jigs?

When I was transferred from the erecting shop to the drawing office – more years ago than I care to remember – the first job I was given was the design of some jigs for the large-engine machine shops. I disliked the work intensely and was greatly relieved when some engine design work came my way after some weeks, but I did learn the value of jigs and a little of how they were made. This early experience served me well in later life, and not least in my model engineering, so I thought that you might like to have a few examples.

Jigs may be used – or needed – for a number of reasons. The first of these may be no more than the saving of time; a workpiece may have a number of holes in it which need not be accurately placed, but if each has to be marked out the floor-to-floor time could well be excessive. A simple plate-type drilling jig will save much of this, and the decision as to whether to use a jig or not depends only on whether the time taken to make the jig is less or more than the time taken to mark out the number of workpieces. In model engineering circles if one has a single cylinder with two covers it is easy to mark each cover and then to use the covers as jigs for the cylinders; but for a two-cylinder engine it may well be quicker to make a simple jig to drill all four covers *and* the cylinders, with the added advantage that the covers are interchangeable.

Which is the second reason for using jigs. If we sold an engine to a customer overseas there would be some choice language floating around the engine-room if a spare part did not fit! A more interesting use was, perhaps, the case of crankshaft couplings. These were drilled to within a couple of thou reaming size at our works, but the generator might well be made at the other end of the country or even in Germany or Switzerland. Their drilling had to be within reaming tolerance of ours, so that the ground bolts would fit. Further, the generator might not meet up with the engine until both got to their destination, perhaps on the other side of the world. So, having used the jig on our shaft, it was then sent to the generator makers for them to use in drilling their coupling. Thus we could be sure that when assembled it would be possible to put a reamer through both halves of the coupling without difficulty. To be sure that we were not held up for lack of the jig, we had several made from a master jig, used solely for making them.

In that case, the important matter was that the holes in both couplings be in their correct relative position – it would not matter if one hole were

slightly closer to the others as both couplings would be the same. There are many similar cases, but there are situations where it is important that holes be a specific distance apart. In this case the jig is no different, but is made with far greater care, using a jig-borer to position the holes.

Well, few model engineers have jig-borers, but you can come pretty close to it by using the co-ordinate method I am going to describe over the next few pages. Incidentally, the main difference between this method and that of the jig-borer is that the latter does not depend on the accuracy of the feedscrews, the table being positioned by setting the required number of accurate slip-gauges between the table and a fixed stop. A special device is used to ensure that the gauges are squeezed up by the same force each time, and the whole is in a temperature-controlled room to reduce errors from thermal expansion of work, machine, and gauges. Not exactly the average model engineer's conditions!

A final reason for using jigs may be convenience. A casting may well be of such a shape that marking out would be a formidable task. In such a case the work can be set in a box jig or fixture and drilled from that. Such a jig may take time to make, and can be of enormous dimensions, but even when building only a few engines it could be considered worthwhile. An important point here was that the engine designer had to have this in mind in the design stage; some of the odd projections you may see on castings may well be there to support the work in a box jig at the machining stage. (Some of the odd holes that seem to serve no useful purpose on your lathe may be there for the same purpose; there is a $\frac{1}{4}$in. hole in the

Myford dividing plate which puzzled me until I found it was a jig location hole.) Model engineers might well keep this in mind if they are designing their own prototype and also the golden rule of all production – to ensure that there will be adequate reference faces in the design from which dimensions may be taken!

In the next few pages I shall deal with the question of achieving accuracy, for if you *are* going to make a jig it is worth taking a little trouble and getting it just right. Further, there are cases where, even if no jig is to be used, accuracy of setting out is important. Then I will cover the more general area, that of getting holes in the right places; and after that I will give details of some of the more special jigs and fixtures which I have found useful.

Co-ordinate setting out

Clearly if a jig is to be made it should be as accurate as possible, and the normal marking-out methods used by model engineers may not be up to it. I make use of a technique known as co-ordinate setting to ensure accuracy; a system which, in effect, changes the usual dimensions found on a part drawing to one which takes all dimensions from fixed datum lines.

To show how effective this can be a test was made at a technical college some years ago. The guinea-pigs were all apprentice toolmakers who had had some years of experience, and they were supervised by a very skilled instructor. They were asked to set out and then drill and ream six holes on a pitch circle. Four men to use normal marking out with dividers, four to use the co-ordinate marking-out method and then drill from the marks, and the third set of four to use a compound table to mark out *and* drill. Each man did his own

workpiece and the results were then measured using very sophisticated measuring tackle. The results were as shown in the table, all figures being in thousandths of an inch.

Method	Mean Error	Max Error	Min Error
Normal marking out	11.2	26	3
Co-ordinate marking out	8.0	12	6
Co-ordinate drilling	3.3	4.9*	0.8

*This one was a 'rogue' not typical of the whole.

By model engineering standards these were skilled practitioners doing similar work all day long, so it may be assumed that the amateur will be worse rather than better, and we cannot expect to achieve mean positional errors – with normal marking-out methods much less than $\frac{1}{64}$ in. and maximum errors twice this figure. Some of this error will, of course, be due to factors other than marking out; drill wander, for example, as seen at * in the table.

Most model engineers have the equipment necessary to reach the standard of the third group in the table; indeed, many will be able to do better, as the machine used was well past its best and had suffered many years of hard work in the hands of less skilled workers! First, however, a word about marking out. It is not always possible to use the method I am going to describe and it is then necessary to mark out, centre-pop, and drill. However, the co-ordinate method can still be used to improve accuracy, especially if a vernier height gauge is available. Even with a normal scribing block it has advantages. All that is necessary is to convert the dimensions in the same way as for co-ordinate drilling and use these from the datum (base of the scribing block as

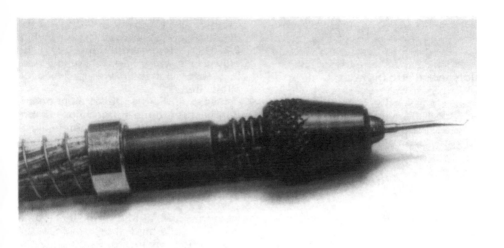

Fig. 2.1 A watchmaker's foret held in a small archimedian drill-driver.

a rule) when marking the lines. The next step, however, may be new to you, that is to use a watchmaker's 'foret', which is simply an extremely hard, tiny spear-pointed drill, very like those once used in fretwork, to make a little dot at the intersections of the lines. I hold mine in an archimedean drill stock – again, like those used in fretwork – and with the aid of a magnifying glass it is possible to align the point exactly (or almost so) at the right spot (see Fig. 2.1). This tiny centre-hole can then be deepened if need be with a punch in the normal way. This technique can bring mean errors well down, as shown in the second line of figures in the table.

Now for the co-ordinate drilling, which is exactly the same as used in industry on a jig-boring machine. All that is needed is a centre lathe with an index on the cross slide and a vertical slide, also with an index. A vertical milling machine will serve equally well – better, in fact, for the fixing of the

work is easier. It goes without saying that the method can be no more accurate than the accuracy of the feedscrews, of course. Consider the example shown in Fig. 2.2 to illustrate the principle. Four holes are required at the corners of a 1-inch square. Never mind their relationship with the edge of the workpiece at present, and concentrate only on their relative positions. The first step is to work out their position relative to two bases or datums (Fig. 2.3) OH horizontal and OV vertical. These can be anywhere which is convenient, but

Fig. 2.2

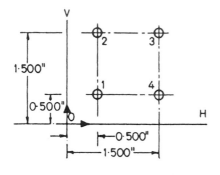

Fig. 2.3

must be at right angles. The holes have been numbered for identification. The position of each hole is now stated in terms of their distance from the point 'O', always giving the horizontal figure first (though, in practice, you would just write them down on the drawing). The table shows these co-ordinates.

Hole No.	Horizontal	Vertical
1	0.500	0.500
2	0.500	1.500
3	1.500	1.500
4	1.500	0.500

Now set up the vertical slide, making sure that it is dead square across the machine. Clamp the piece to the slide with suitable parallel packing behind to allow the drill to go through. Make an arbitrary decision about the position of the point 'O', and make a note of the index readings. Let us suppose that this is where both indexes read zero (we will come to the method of positioning the holes relative to the edge of the work-piece in a moment). Now, a very important note; *all readings of the index, including that at zero, must be made with the handles rotating the same way.* I have got into the habit of using a

rotation which moves the slides in the direction of the arrows shown in Fig. 2.3. This is *most* important, being the only way in which backlash, or any slack in feedscrew handle washers, can be eliminated.

Assume now that the drilling operation is to be done in the lathe. Grip a Slocombe drill either in the three-jaw or in an accurate drill chuck held in the headstock taper. Wind the cross slide forward 0.500 in. and the vertical slide up 0.500 in. Advance the saddle and drill deeply at an appropriate spindle speed as fast as you can for this size. Withdraw the saddle. Wind back both slides a little and then advance the cross slide, again to 0.500 and the vertical to 1.500. Drill as before. That is hole No. 2. Wind back a trifle, and then advance horizontal 1.500 and vertical 1.500. Drill No 3. Wind back, and then advance H-1.500, V-0.500 and drill hole No. 4. Now replace the Slocombe with a No. 14 drill and repeat the whole operation, drilling right through. Don't be tempted to slack up on the care in setting the indexes just because you have a deep centre to drill to! Finally, replace the drill with a reamer, and at much slower speed and plenty of oil, repeat the whole. You can check your accuracy by making a second piece exactly the same, and checking the holes in the two pieces one on top of the other *in any position*. With reasonable luck you should be able to push pieces of 0.186 in. diameter rod through all four holes at once.

A little explanation of the instruction to wind back between co-ordinate settings. This is simply to avoid errors which might result from vibration causing movement of the slides, as in this example adjacent holes each have one co-ordinate the same.

If the holes are to be drilled in the drilling machine then, of course, the workpiece can be removed after using the Slocombe. It is not always convenient to drill right through on a vertical slide and with larger drills the strain on the saddle feed rack – and on the wrist – can be high. A little accuracy is sacrificed. If the position only of the holes is needed, without any drilling, then fit a marking centre in the headstock and simply press the work up against it. Clearly if the milling machine is used, with a vertical quill, the operation is similar, but has the advantage that the work can be positioned on the flat table instead of on the vertical slide.

Location of holes from edges
It is seldom that holes can be planted any old where on the workpiece. They must usually be located in relation to edges or centres. Consider the first case – location from two edges – and

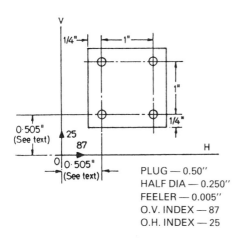

PLUG — 0.50″
HALF DIA — 0.250″
FEELER — 0.005″
O.V. INDEX — 87
O.H. INDEX — 25

Fig. 2.4

assume the same geometry of holes, see Fig. 2.4. Assume also that your three-jaw is reasonably accurate, or can be brought within 0.001 in. by using pieces of paper under one or more jaws. Grip a piece of 0.500 in. ground stock (any diameter will do so long as it is accurate to diameter, true, and stiff; not less than $\frac{1}{8}$ in.). Let it project far enough to reach the workpiece past any clamps and check its truth. Set the workpiece on the vertical slide so that the sides adjacent to the proposed baselines are truly horizontal and vertical. (It is assumed that the two sides are truly square to start with; if not, make them so.) Measure the stock in the chuck and write down its diameter (see page 139). Divide it by two. Select the best of your feeler gauges and write down its thickness, say, 0.005 in.

Now advance the cross slide in the direction of the arrow OH until the feeler slides sweetly between the workpiece and the test plug in the chuck. The edge of the work is now exactly the half diameter of the plug plus the feeler thickness from the lathe centre – 0.255 in. in this case, so that the base CV is 0.255 + 0.250 = 0.505 in. from the dimension line through hole No. 1. Make a note of the index reading, say 87 (or adjust the index to zero if this can be done). Repeat in the vertical plane, thus locating OH relative to hole 1 again. Note the index reading – say 25. It helps to note the readings on the drawing as shown in Fig. 2.2, but note that the CV index refers to the position of CV and is used in traversing along OH. You can now log the co-ordinates and the appropriate index readings, or mark them on the drawing as you prefer. The log appears as follows, assuming the screws have 10 tpi threads, with 100 divisions on the index.

23

| | Horizontal (87) | | Vertical (25) | |
Hole No	Co-ord	Index	Co-ord	Index
1	0.505	5/93	0.505	5/30
		(.505+87)		(.505+25)
2	0.505	5/93	1.505	15/30
3	1.505	15/93	1.505	15/30
4	1.505	15/93	0.505	5/30

where the figure 15/ means fifteen turns of the screw and the 93 means turn further until this index is reached, and so on.

Frankly I find it easier to mark it up on the drawing, or even make a special sketch, rather than use a log, but whichever method is used it is essential to write it all down!

Location from given centre lines
This is nowhere near so easy and I always prefer to re-dimension the work so that location can be made from datum edges. However it is often not possible to do this, and the following method is then adopted. Chuck a piece of $\frac{1}{2}$ in. carbon tool steel, marking the position of the No. 1 jaw so that the device can be accurately set in future. Turn a point after the style of Fig. 2.5 to project far enough to clear clamps etc. Get the point reasonably sharp. Remove

from the chuck and harden. Then temper by heating the cylindrical part until the point reaches a straw colour, and quench. Return to the chuck, but resist the temptation to polish the taper – you want as little reflection as possible. If the point is now blunt, sharpen by running the lathe fast and touch it up by application of an oilstone. If the point does not run true it may be necessary to pack under the chuck jaws with paper, but the odd thou will not matter.

In use, it is only necessary to advance the work to the point until, using a magnifier, it is seen that the point is exactly at the intersection of the centre lines. Do not allow the point to *touch* the work, as there is a risk of breaking it. Do not forget that the slides must always be wound in the direction of your OH and OV arrows. Having made a note of the index readings it is then a simple matter to establish the position of suitable bases. These *can* run through the centre you have marked up to, but it is always preferable to have bases outside the workpiece itself – then all co-ordinates are forward ones, and there is less risk of forgetfully winding the slide the wrong way.

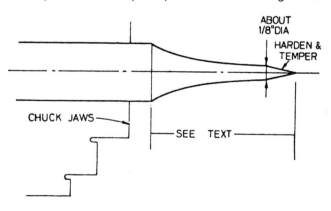

ABOUT 1/8"DIA

HARDEN & TEMPER

CHUCK JAWS

SEE TEXT

Fig. 2.5 A 'dotting' marking punch for use in the chuck.

If you have a device for setting lathe tools to centre-height (see page 58), this can be used for setting in the vertical plane provided it is accurate. My own was specifically designed for this purpose, and can also be used for marking out work held in the lathe. It is, in fact, a fixed height gauge and has hard chisel points similar to those used on vernier height gauges.

Setting to an existing hole
If the hole is fairly large – say a cylinder bore – the most accurate way is to set up a dial indicator in the chuck and adjust the position of the slides (always traversing in the OH and CV directions) until it is found to be concentric with the spindle on pulling the belt round by hand. The same method may be used for smaller holes if they are of such a size that a plug can be fitted in them without shake, the DTI being traversed round the plug instead of within the bore. In cases where split thousandth accuracy is not necessary, a very simple method is to fit a true centre in the headstock taper and advance the saddle so that the end of the centre is within the hole. The position of the sides is then adjusted until the centre touches the edge of the hole all round. Care must be taken, as it is quite possible to shift the workpiece in its clamps if the saddle is pushed too hard towards the

headstock, but if you can get a square up to the workpiece at the same time, against some reference face, you *can* take advantage of this fact. Simply set both slide indexes to zero, roughly position the workpiece and lightly clamp up. Advance until the centre enters the hole and then carefully bring the saddle forward until it presses the centre into the hole sufficiently to move the work. Square up, and carefully tighten the clamps. Without moving the saddle, check the slide handwheels, winding in the correct direction. You will almost certainly find that the operation has moved the slides within the slack in the slide nuts. Note the reading on the indexes, and proceed as before.

More complicated hole arrangements
A very simple arrangement of holes has been selected to illustrate the principle. Clearly the operation will not be much different if the holes were six in number, arranged in a rectangle, as on a valve-chest (in which connection I always drill cylinder, chest, and chest cover, at different times, using this method, and have never had to draw a hole since I started; and the covers usually fit either way round!). With more complicated arrangements, however, some recourse to geometry is necessary, and a few examples follow. Fig. 2.8 will refresh your memory on the

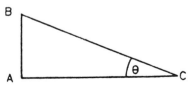

$$\frac{AB}{AC} = TAN\ \theta \quad \Big| \quad AB = \sqrt{BC^2 - AC^2}$$

$$\frac{AB}{BC} = SIN\ \theta \quad \Big| \quad BC = \sqrt{AB^2 + AC^2}$$

$$\frac{AC}{BC} = COS\ \theta \quad \Big| \quad AC = \sqrt{BC^2 - AB^2}$$

Fig. 2.6   Triangular and angular relationships.

25

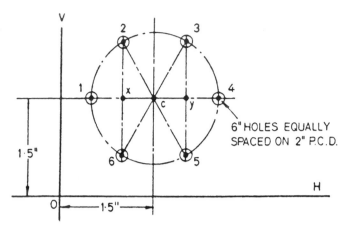

Fig. 2.7
Holes in a circle,
equally spaced.

6" HOLES EQUALLY
SPACED ON 2" P.C.D.

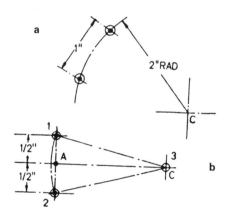

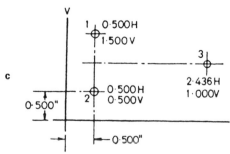

Fig. 2.8 Holes in a circle to a pitch dimension.

basic geometry required – the necessary tables of values, as well as of squares and square roots, will be found in *The Model Engineer's Handbook* (Tubal Cain, published by Nexus Special Interests Ltd.), and comprehensive tables can be bought for a small outlay from bookshops. It will be seen that there are several ways of finding the length of any size of a triangle if either the angle or the other lengths are known, and if the arrangement of holes is split up into a series of such triangles the dimensions can be rearranged into co-ordinates as for a rectangle.

Holes in a circle
Fig. 2.7 shows six equally spaced holes on a 2 in. diameter circle. Select a position for the bases OH and OV outside the circle at a convenient distance from the centre, say 1.500 in. The co-ordinates of holes 1 and 4 are easy, 1 being 0.500H, 1.500V and 4 being 2.500H, 1.50OV. Now take hole 2. We need the distances 2-x and x-C. The triangle is x2C and the angle is 60 degrees. Length 2-C is known, as 1 in., the radius of the circle. Hence (from Fig. 2.6) 2-x = 2-c Sin 60

26

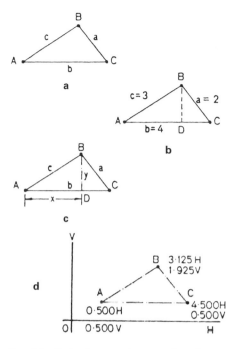

Fig. 2.9   Holes in an unknown triangle.

= 1 x Sin 60 or 0.866 in. Length x-c is given by z-C Cos 60 = 1 x 0.500 = 0.500.

Thus the co-ordinates of hole 2 are 1.000H, 2.366V(1.5 + 0.866). Note that the other lengths we need are the same as those already found in this case 2-x = x-6 = 3-7 = y-5, and x-C = C-y. The table of co-ordinates then reads:

Hole No.	Horizontal	Vertical
1	0.500	1.500
2	1.000	2.366
3	2.000	2.366
4	2.500	1.500
5	2.000	0.634
6	1.000	0.634
Point C	1.500	1.500

From this example it will be seen that it is possible to locate all the holes

in the cylinder head of an engine within a few thousandths of an inch. The method is more accurate than is possible with the normal dividing disc.

Holes dimensioned to radius and pitch
Fig. 2.8a illustrates this case. The first step is to redraw the arrangement as in Fig. 2.8b with the holes 1,2, equidistant from the centre line AC which runs through the hole C. The angle 1c2 is not known so that the other relationships shown on Fig. 2.6 are used. Note that we have altered the isosceles triangle to the two right-angled triangles to make the calculation easier.

From Fig. 2.6

$$AC^2 = 1C^2 - 1A^2$$
$$= 4 - 0.25 = 3.75$$
$$AC = \sqrt{3.75} = 1.936 \text{ in.}$$

The co-ordinates of the three holes are now shown on Fig. 2.8c, the two bases being $\frac{1}{2}$ in. removed from hole no 2. These co-ordinates assume that the workpiece is set up in the machine with line AC horizontal.

Holes in a triangle – angles not known
This is a more difficult case, for if the triangle cannot be rendered right-angular and if no angles are known, the relationships in Fig. 2.6 cannot be used. Fig. 2.9 shows such a case. Denote the lengths of the lines by small letters corresponding to the opposite angle e.g a opposite A, b opposite B etc. Then the relationship is given by:

$$\text{Cos A} = \frac{b^2 + c^2 - a^2}{2bc}$$

and $$\text{Cos B} = \frac{a^2 + c^2 - b^2}{2ac}$$ *Etc.*

Putting in the figures will give the angles A and B. We can then use these angles to find the lengths x and y in Fig. 2.9b by writing, from Fig. 2.6:

$$y = c.\text{Sin A (OR } a.\text{SIN B)}$$
$$x = c.\text{Cos A}$$

Note that it is not *necessary* to know the angle B, but it is useful as a check on the working. The length of y should come out the same in each case.

A worked example is taken from Fig. 2.9c.

$$\text{Cos A} = \frac{16 + 9 - 4}{2 \times 12} = \frac{21}{24} = 0.8750$$

From, tables,

A = 28 deg. 57 min.

BD = y = 3 × 0.4841 = 1.452 in.

AD = x = 3 × 0.8750 (from above)

= 2.625

Fig. 2.9d shows this triangle rendered into co-ordinates. You may care to check this – easily done by drawing the figure to scale.

Useful special case

A very useful geometrical truth is shown in Fig. 2.10. This shows a number of points, 1, 2, 3, disposed around a pitch circle, of which AB is the diameter. In this case, the angles A1B, A2B, A3B, etc., are *all right angles*. The theorem is stated in the geometry books in the words "the angle inscribed in a circle on the diameter is a right angle". This property can be made use of in setting out co-ordinates in some circumstances, and saves time in finding angles, especially when the points are not equally spaced.

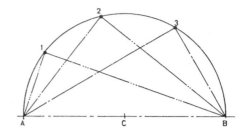

Fig. 2.10   Angles in a semi-circle.

Useful accessories

Enough has been said of the method to enable you to develop it further for yourself. It will be appreciated that the principle is exactly that used in jig boring, except that precision slip gauges are employed for measuring the co-ordinates on the machine, rather than relying on the feedscrew indexes. Computer-controlled machine tools use the same principle, the computer being used to calculate the co-ordinates. A few words on useful accessories may help.

Fig. 2.11 shows a scribing centre which may be used for any marking out done by co-ordinates. It simply comprises a very hard steel scribing point held in a Morse taper shank in the headstock, with a fairly strong spring behind. In use, the workpiece is set to the desired co-ordinate in one plane, and then brought up to the scriber

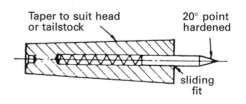

Taper to suit head or tailstock

20° point hardened

sliding fit

Fig. 2.11   The spring centre.

28

which is depressed until it only just emerges from the shank. The work-piece is then traversed across in the other plane over the necessary range of co-ordinates. In the case of Fig. 2.4, for example, one could drill the four holes first, and then mark out for the outline of the square. A similar method is used when marking out for cylinder ports in those cases where co-ordinate milling is not to be applied. (This requires the end mill or slot drill to cut to an exact width and this, in turn, depends on the run-out of the chuck or arbor used.)

Fig. 2.12 shows a device used when a number of components had to be machined with the same vertical co-ordinates. The vertical slide is set to each co-ordinate in turn and a gauge-block made (shown shaded). A similar block is then made for each vertical setting. There is then no need to use the index – the appropriate block is fitted

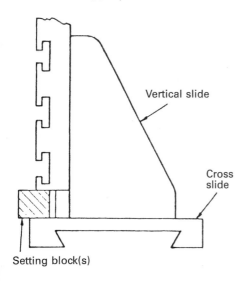

Vertical slide

Cross slide

Setting block(s)

Fig. 2.12 Use of the co-ordinate setting block gauge.

and the slide screwed down until it can go no further. It is no more accurate, but saves time and avoids mistakes in setting. I use a similar device on the table of the milling machine which has adjustable stops. These are set first and locked, and then gauge bars interposed for the various table positions.

I have taken rather a long time over this section, because those unaccus-tomed to what was called geometry at school may not be familiar with the principles, and those who are may not appreciate the possibilities. When working on models of my own design I now tend to dimension all holes from datum lines and increasingly use the miller instead of the drilling machine – it is, in effect, a drill with a compound table. However, as it is not always convenient to pack up the work, as a rule I start all the holes with the appro-priate size of Slocombe drill. I can then move to the drilling machine to finish the job. When making plate jigs, how-ever, I always drill right through and ream on the miller.

In conclusion, to repeat two impor-tant points. First, get into the habit of stating the *horizontal* co-ordinate first when writing them down. Secondly, always traverse feedscrews in the same direction when coming up to the mark, and if you overshoot always go back at least half a turn before going forward again.

Location jigs

I do not propose to describe jigs for every purpose – it would take far too long, and in any case, part of the fun lies in thinking out your own solution. But I will illustrate one or two devices which may help to suggest others.

My first example is that of drilling a cross-hole in a spindle. Everyone knows

29

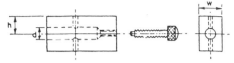

Fig. 2.13a   Cross-drilling jig with end stop.

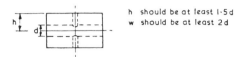

h should be at least 1·5d
w should be at least 2d

Fig. 2.13b   Straight-through jig.

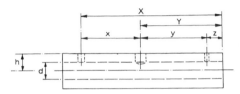

Fig. 2.13c Cross-drilling jig for spacing holes.

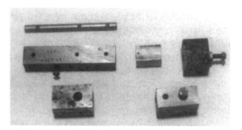

Fig. 2.13d   A collection of small cross-drilling jigs with a sample workpiece.

how to do this, but perhaps not all realise how greatly it can be extended. Two forms of the jig are shown in Fig. 2.13a and b, the only difference being that one has an end-stop and the other not. The former is needed when the hole must be positioned in relation to one end, but I use the latter when the rod needs subsequent machining operations – a case I will come to in a moment. The main problem is to get

the two holes truly in line. Do it this way. The piece of square or rectangular material is held in the machine vice mounted on the vertical slide, and one hole is drilled. If it isn't convenient – or possible – to ream, use an undersize drill first and then open out to correct sizes. Without shifting the vertical slide the work is now rotated through 90 degrees (set it dead square in the first place) and the second hole drilled. You may have to adjust the cross slide, but provided you have not interfered with the vertical setting the two holes must be in line. Drill and tap for the stop-screw if you need one. I think, too, it pays to enlarge the one end of the cross-hole for the drill to break through into, but there are cases where you can't do this. If you do, you must mark which side is top – as well as the sizes of the holes. Harden it? If you like; it depends on how much use it is likely to get, but it is true that you only need one drill to walk a little and you may spoil a soft jig.

If a number of cross-holes are to be drilled in the workpiece the arrangement shown in Fig. 2.13c may be used, although I will describe an alternative later. To make this type, the longitudinal hole must first be drilled (and reamed if desired) in the four-jaw, taking more than usual care over clearing chips so that the hole is straight. The cross-holes are then drilled using the method already described. It is clearly necessary to take care in setting up for the centre lines, both in the four-jaw and when mounting in the vertical slide. The dimensions 'x', 'y', and 'z' are set up using the cross-slide index and are best recalculated as co-ordinates from the reference face of the jig as shown at 'X' and 'Y'. I use this fixture when making handrail stanchions, the holes in the stock

30

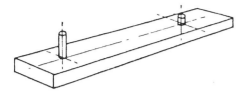

Fig. 2.14 *Hole-spacing fixture for use with Fig. 2.13b.*

being drilled before any turning is done.

This method does present some problems as I have already suggested, and you may, for example, meet some difficulty in extracting the workpiece after drilling. I tend to use this type of jig only when I have a fairly large number of similar workpieces to make. In most cases my practice is to drill the holes in the plain stock, with an excess of length, first, and then to do all the turning operations. Step 1 is to drill one end of the stock, using the jig at Fig. 2.13a. I then make a plate-jig (but see later also) as Fig. 2.14.

The already-drilled stock is slipped over the long peg in the plate, the drilling jig is located by the short peg in the second hole, and the stock drilled. Provided the fit of the pegs is good, and you have taken reasonable care to align the drill to the work the two holes must be dead in line and at the correct centres. Fig. 2.15 shows the thing in use, and Fig. 2.16 a finished rod. Note that in this case it is important *not* to enlarge the bottom half of the hole in the drilling jig.

Fig. 2.15 *A universal hole-spacing jig in use.*

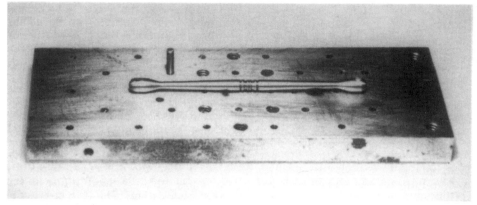

Fig. 2.16   Link turned from the blank in Fig. 2.15.

In these photos, you will observe that the base is no more than a lot of holes surrounded by metal. This makes a good universal jig. This particular piece came to me with some junk at an auction, and what it was for I don't know. Fortuitously the holes were so spaced that I could use it for centres advancing by $\frac{1}{2}$ in. one way, and by $\frac{9}{16}$ in. another, with a few centres at the $\frac{1}{4}$ in. interval, but it gave me the idea that a fully universal jig could easily be made, Fig. 2.17. You can, of course, space the holes, and have as many of them to suit

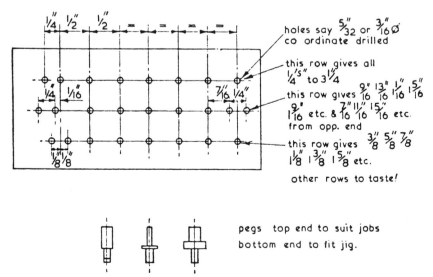

Fig. 2.17   Suggested dimensions for a universal plate jig.

your taste, but if odd lengths are required it is usually better to make a 'special'. In this case the pegs are stepped, so that one end fits the base jig and the other the hole in the drilling jig. These pegs may have to be made specially for each job, but that doesn't take long and the device saves hours of time.

An alternative method is to make a cross-drilling jig long enough to accept the blank and drill all the holes with this. It is quicker, as it involves less setting up, but it is a little more difficult to make as you need a long, parallel hole instead of a relatively short one. I have one or two such, but only for components needed in quantities of more than half-a-dozen.

Eccentric rod jig

A special case is shown in Fig. 2.18. It is, as everyone knows, imperative that the eccentric rods of any Stephenson link motion be of the same length, otherwise it will be impossible to time ahead

and astern events correctly. The procedure is to machine the bore of the eccentric strap to size, and make and fit the eccentric rod to it. The strap is fitted to the peg 'a' and the end of the rod into the 'gubbins' – 'b'; this latter is no more than a drilling block with a slot in it. A peg is slipped into the base block and into the lower part of the gubbins to locate the eccentric rod sideways, as at 'c'. This arrangement has proved very successful with small rods, especially those made in one piece, which cannot be fine-adjusted by filing the foot of the rod.

For large rods I use the jig shown in Fig. 2.19. The principle is the same, but in this case the eye end of the rod is drilled and (if need be) bored in the lathe. The hole for the locating peg, and the setting centre, are set out by the co-ordinate drilling method. The peg is made a good fit on the eccentric bore and a hard push fit in the base. (You may wish to take it out and use for a different job later.) The base is set true

Fig. 2.18 An eccentric-rod drilling jig.

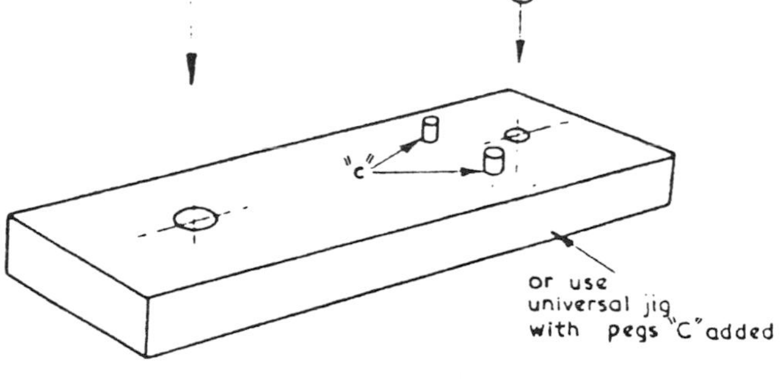

"a"

"b"

"c"

or use
universal jig
with pegs "C" added

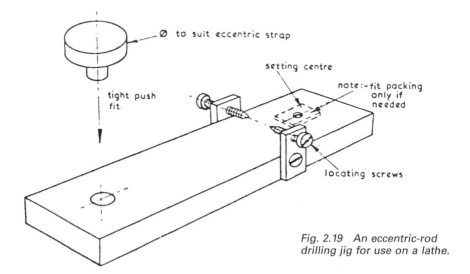

Ø to suit eccentric strap

setting centre

note:-fit packing only if needed

tight push fit.

locating screws

Fig. 2.19 An eccentric-rod drilling jig for use on a lathe.

on the faceplate to the locating centre, and a balance-weight fitted so that the lathe can be run fast enough. The eccentric is located sideways by adjusting the two screws (or you could simply use pegs) which I find, if nipped up, will be sufficient to hold the job without clamping. The eccentric is, of course, tightened onto the peg, and if not really tight a piece of cigarette paper put under the cap.

An almost identical jig is used for dealing with any connecting rods short enough to be swung in the lathe, and I use this even if I have only one connecting rod to make. It has the great advantage that it ensures that the bores of the small and large ends are truly in line, in both planes. Any twist will, of course, cause binding, while misalignment in the other plane will cause sideplay at the crosshead or on the gudgeon pin in an IC engine.

A brazing jig
The next device – Fig. 2.20 – is a further extension of the principle, designed to facilitate the brazing-up of a set of six drag-links. These were made by brazing bosses to the ends of a steel strip. In this case the centre-distances were not critical and alignment error could have been corrected by slightly twisting the links, but a jig was used solely for convenience in holding the parts while brazing. The sketch is self-explanatory for the most part, the jig being cut from a solid piece with a slot drill; as the stock was in the vertical slide anyway, the holes were co-ordinate drilled before tapping. Why use screws instead of pegs? Experience showed that molten flux could solidify over pegs, and make it difficult to remove the finished job, so stainless steel screws were used instead. Further, to ensure that the work was not brazed to the jig by accident the latter was heated before use until a good film of oxide was formed all over, too heavy for the flux to reduce. One point has to be watched; if the jig is made very heavy then you may waste a

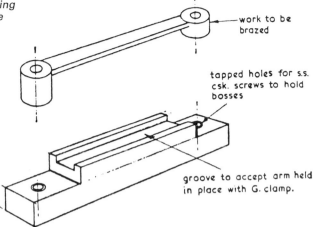

Fig. 2.20 Jig for holding parts while brazing the bosses.

work to be brazed

tapped holes for s.s. csk. screws to hold bosses

groove to accept arm held in place with G. clamp.

lot of gas as the heat tends to be conducted away, but if too light a section is used then it may distort and fail to serve its purpose. The use of a soft mild steel, preferably black stock, will reduce the latter risk, but the end 'steps' should not be too thin.

A cylinder boring fixture and alignment jig

I now propose to describe a fairly elaborate jig set-up as a bit of a contrast to what has gone before, and to illustrate how a jig may be needed for a one-off job to ensure alignment of an assembly. This bit of jiggery was designed to cope with the building of a marine engine in which the cylinder assembly is supported on columns from the crank-bed (see Fig. 2.21). A little reflection showed that there were considerable possibilities of misalignment if the work was carried out in the usual way by marking out. Co-ordinate drilling could have reduced these errors, but not all. The columns must be correctly set out in relation not only to

the bed centre line but also to the main bearings axially. Similarly, the column holes in the cylinder block must be in correct relationship to the two bores, and these bores to each other.

If the old-fashioned device of making the columns a slack fit so that the cylinder could be adjusted had been adopted, there still remained the need for the cylinder bores to be aligned and spaced in relation to the main bearings. Had the engine been larger the problem might have been a little easier, as some end-float could (and in any case should) have been allowed at the crosshead pins, but in a small engine it was quite impossible to leave enough float to accommodate the probable, let alone the possible maximum misalignment. The solution adopted was to design a plate-jig for drilling both bed and cylinder block, and then to use this jig to make a second one to hold the cylinders for boring on the faceplate. This reduced the probability of errors due to machining to those of the dimensional accuracy of the crankshaft,

35

Fig. 2.21 *A small model launch engine. The design can present alignment problems.*

as the main-bearing thrust faces would, in any case, have to be fitted in the usual way. This fitting operation would, it was expected, absorb any error in the shaft, and this proved to be the case – an end-float of three thou, at the cross-heads absorbed all other errors.

Fig. 2.22 shows the outline of the two

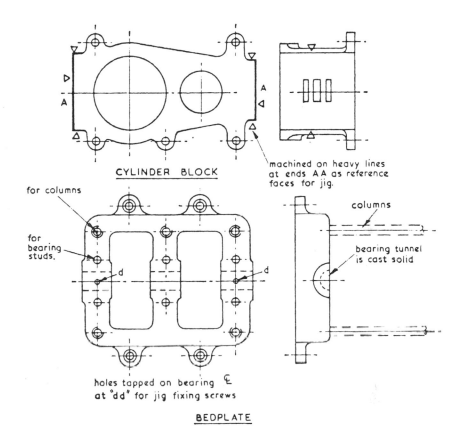

CYLINDER BLOCK

machined on heavy lines
at ends A A as reference
faces for jig.

for columns

for
bearing
studs.

columns

bearing tunnel
is cast solid

holes tapped on bearing ₵
at "dd" for jig fixing screws

BEDPLATE

Fig. 2.22 Cylinder and bedplate of a compound launch engine.

components to be machined. The two faces AA on the cylinders were used as reference planes, and after machining the top and bottom of the block these faces were milled on the vertical slide, the only marking out that was done being the longitudinal centre line and witness lines for the faces AA. Incidentally, these were set out in relation to the column bosses, not the bores, as there was ample metal in the latter for machining. So far as the bed was concerned this also was first machined top and bottom and the longitudinal centre line marked out, as well as that for the centre main bearing. This was then set up in the vertical slide and a hole drilled and tapped in the centre of the two outer main bearings, these holes being determined by co-ordinates on the cross slide. These holes were to hold the drilling jig to the bed, as clamping would be difficult.

Fig. 2.23 is the drilling jig, made of plate about $\frac{1}{8}$ in. thick, with the two reference fixtures at the ends of the

37

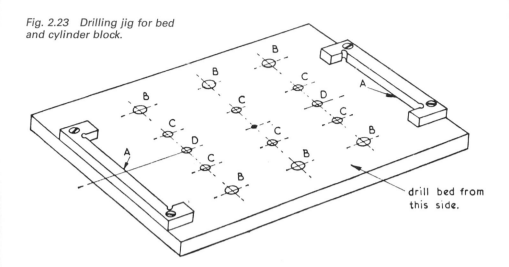

Fig. 2.23 Drilling jig for bed and cylinder block.

drill bed from this side.

same material. Again, the only marking out that was done were the two centre lines, and even this was not really necessary. The holes B are tapping size for the columns, and it will be observed that there are six of these, though only five columns, this being a requirement for the jig mentioned later. Holes C are tapping size for the main-bearing studs – if a jig is to be made these might as well be jigged up also. (The main-bearing caps can be drilled in a simple box jig or by co-ordinates in the lathe.) At D are the holes for the screws to hold the jig to the bedplate. The reference plates AA are made to fit the reference faces on the cylinder block, and the only measurement that needs to be done on the jig is the setting of these plates – all holes were drilled by co-ordinates. Even here, a slight error is not impor-tant, the only effect being to put the slide-valve faces a trifle out; a tolerance of ± 10 thou. would not hurt.

The jig having been made, it was marked with yellow paint round the

surplus hole B to remind that this was not needed except on the boring jig, and also marked to show from which side of the jig each component was to be drilled. Incidentally I should have mentioned that *all* holes were drilled in two stages to ensure that they were the correct size.

The making of the jig took surpris-ingly little time, and drilling of the workpieces only a few minutes, and it is likely that it would, in fact, have taken longer to mark out the two parts and drill in the conventional way; the cylin-der would not have been too easy to hold for marking out.

Fig. 2.24 is the boring fixture; this is made of more substantial material, and selected to be of even thickness. Had such not been available I should have used a piece of thick frame-steel and riveted two thick stiffeners each side, checking with a micrometer to see that the overall thickness was uniform with-in a few thou. To make the fixture, the two main centre lines were first set out,

38

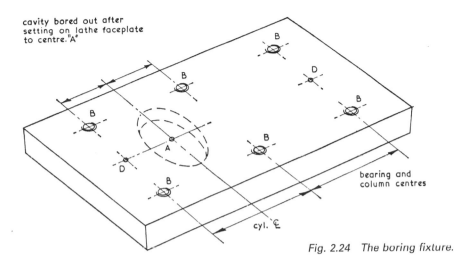

cavity bored out after
setting on lathe faceplate
to centre."A"

Fig. 2.24 The boring fixture.

taking care that they were square to each other, and the job then clamped up to the vertical slide (left there from making the other jig) and the centres A and DD set out and marked by the co-ordinated method. 'A' was lightly centred with a tiny Slocombe. The drilling jig was then set over the faces and the centres DD observed through the corresponding holes in it – also marked D in Fig. 2.23. When truly centred – and this is a more accurate procedure than many might imagine, as the eye has considerable critical acuity when any misalignment is observed – the two were clamped together and the tapping holes at BB were drilled. (I should have mentioned that the drilling jig holes are tapping size, and the clearance holes in the cylinder block opened out afterwards.) The fixture was then set up on the lathe faceplate and accurately centred to the point A, after which the cavity was bored out to allow the boring bar to emerge from the cylinder bore. This cavity was in fact bored to a good

finish, so that it could be used for setting-up again should the same job turn up a second time.

The cylinder block was then attached to the jig with six screws; in fact four would have been enough to take the cutting forces but by using six any differences between screw and hole were averaged out, making the overall setting pretty accurate. The first bore was roughed and finished, and the casting then removed and reset with the other bore presented to the tool. As the jig had not been touched, this was bound to be correctly aligned. Fig. 2.25 shows the set-up, with the drilling jig alongside, and Fig. 2.26 boring in progress. This procedure has taken a lot of words to describe but in fact it was remarkably rapid throughout. It will be noted that there was very little marking out to be done – a time-consuming operation even with a vernier height gauge – and that the biggest risk of normal procedure was avoided almost entirely, that of the drill point wandering sideways. The slowest part

Fig. 2.25 The cylinder set up for boring (the drilling jig is on the left).

Fig. 2.26 Boring in progress. The boring bar is damped as shown in Fig. 4.28 (p. 138).

of the work was the conversion and subsequent checking of the dimensions to decimals.

I could extend this section considerably, but I hope that the examples given will give you some idea of the possibilities. There are, of course, many books dealing with the subject for production engineers, many of them vast tomes. Such complexity is not necessary for the model engineer (and some would say is overdone in industry, too) but I do know some who go to a great deal more trouble than I do; and why not, for if you like jigs it is great fun! I believe, though, that for most applications the simpler the device the better, provided it adheres to the basic principles of working from a definite datum line or plane. One final point before going on to consider a few specialist applications. When you have made the jig – or any such accessory – mark it for what it is before you put it away! You never know when you may need it again and an unmarked jig can be a snare and a trap, for it just *may* be a special one you have forgotten about, made ten thou over the design size!

A bearing boring fixture
(*Adapted from an article in the* S.I.M.E.C. Magazine)
Large multi-cylinder engines nowadays have the bearing housings in the bedplate line-bored and these take semicircular bearing shells, which may sometimes also be line-bored in situ. The compound, triple, and quadruple expansion steam engines of earlier days, however, had rectangular ways for the brasses, as these could be machined on the planer when the rest of the work on that part was being done. There is much to be said for using this construction on a model, for line boring is not practicable at scale sizes. The recesses in the bed can be milled with a slot drill and (provided the usual precautions are taken) are bound to be in line both vertically and horizontally.

If, however, the brasses are then machined one at a time there may well be considerable misalignment when the shaft is offered up. The little fixture shown in Fig. 2.27 will ensure that the relative positions of the bore and the alignment faces are held within close limits. Naturally the dimensions will differ depending on the size of the brasses you are making; those shown correspond to the Stuart Turner triple expansion engine.

The first step is to machine the rough brasses dead square all over – they will, of course, be made from two pieces soldered together in the usual way. This having been done the recesses in the sides, bottoms and tops of all four must be machined to identical dimensions. Whether done on the vertical slide or a milling machine the procedure is the same – work to co-ordinates, and take great care with the setting up, especially the squareness of the vice.

The fixture shown was made from aluminium alloy, simply because I had a chunk the right size handy, but anything would do: indeed, for only four off, even lignum vitae. The rectangular slot is machined very carefully to be a slide fit to the brasses, and two lines scribed across to correspond to the centre of the bearing. The fixture is thicker than the width of the slots in the bearing brasses, to allow for a recess at the back to clear the point of the boring tool, but you could use packing behind to achieve the same result, in which case the fixture need be only $\frac{7}{16}$ in. thick (for this engine, that is). The little clamping bar has one fixing hole slotted so that it

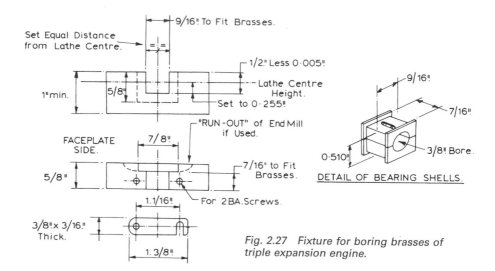

Set Equal Distance from Lathe Centre.

9/16" To Fit Brasses.

1"min.

5/8"

1/2." Less 0·005."

Lathe Centre Height.

Set to 0·255"

"RUN-OUT" of End Mill if Used.

FACEPLATE SIDE.

7/8"

5/8"

7/16" to Fit Brasses.

1.1/16"

For 2 BA. Screws.

3/8" x 3/16." Thick.

1. 3/8"

9/16"

7/16"

0·510"

3/8" Bore.

DETAIL OF BEARING SHELLS.

Fig. 2.27 *Fixture for boring brasses of triple expansion engine.*

is only a case of loosening the screws to change workpieces – less risk of one getting lost in the swarf tray! The edges of the slot should be bevelled so that the brasses don't bind here.

To set it up on the faceplate, clamp up more or less in position and then use your centre-height gauge (or scribing block if you haven't one of those useful accessories) to set the centre lines of the fixture truly at centre-height. Alternatively, if you have a vernier height gauge you can work from the sides of the slot. Once true, clamp up securely, and check again to make sure that the job hasn't shifted when you tightened the nuts. Before using it, mark each of the brasses on one side – say letter A for aft – and make sure when assembling the fixture that this faces the tailstock; when fitted to the bed all should face aft, so that they are all the same way round. Make sure that there is *absolutely no* dirt either in the slot or on the recesses of the brasses, for even a thou displacement

will put the bore out of line by the same amount. Drilling and boring follow the usual procedure: I started by drilling, bored out to reaming size, and finished with a reamer. Fig. 2.28 shows the fixture in action – note the balance weight, which enabled boring to be done at about 600 rpm.

Fixture for machining double eccentrics
All reversing engines using Stephenson gear need a pair of eccentrics providing the same angle of advance ahead and astern. Personally I prefer to use two separate eccentrics, so that each can be set individually, but this is not always possible. Several methods have been published in books over the last 90 or 100 years. The fixture described here was made to deal with a set of such eccentrics in which the angle of advance differed between the HP and those for the IP and LP cylinders. They were about $\frac{3}{4}$ in. diameter with a throw of $\frac{5}{32}$ in., the shaft being $\frac{3}{8}$ in. diameter, but the idea can be used for any size

Fig. 2.28 *The bearing boring fixture in use.*

within reason; the limit is the size of your chuck.

Fig. 2.29 shows the device, and shows that the angles of advance were 120° and 150°. Make the arbor or mandrel first – ground stock will save time on the face of it, but I always prefer to turn the top diameter when doing chuck work. The spigot which fits inside the main body can be reduced by about a thou with advantage. The body should be as large as will enter the

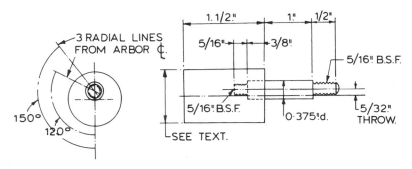

Fig. 2.29 *Fixture for machining twin eccentrics (can also be used for single eccentrics).*

hole in your three-jaw chuck, as it must be capable of being gripped by the full length of the jaws if at all possible. Face both ends and then scribe two lines exactly at centre-height and at right angles on one end. With one of these lines vertical, set off the throw of the eccentric, marked with a short line.

Now transfer to the four-jaw and set this throw dead true *but* with the long centre line aligned with one of the chuck jaws. To do this, apply a spirit level to the jaw and ensure that the line also is horizontal using a scribing block or your centre-height gauge. Once set, tighten the jaws and drill, bore and tap the hole. Now for the setting lines. Mark the original centre line in some way and set this horizontal and to the front of the machine. This need not be exact. Rotate the chuck in the normal turning direction (i.e. top towards you) until the next jaw is nearly horizontal. Set a 30° protractor gauge on top of this jaw and rotate still further until a spirit level set thereon shows the top of the gauge to be horizontal. Check the level both ways round. The jaw is now 30° below the horizontal. Scribe a line across the face of the body, from the hole towards you and mark it temporarily '120'. Repeat this process with the protractor gauge set at 60° and mark this line 150. Naturally, these angles will be set to suit the engine you are making. If you have no protractor gauge you will have to set out the angles on a piece of sheet steel or brass and file this carefully to shape, but for the angles shown an ordinary draughtsman's set-square will do. The fixture can now be removed from the chuck and the arbor screwed in with some *Loctite* retaining compound on the threads. Mark the angles permanently.

The fixture is used in the three-jaw chuck. There is no real need to set any of the lines on the face in line with a chuck jaw but I usually try to get the zero line aligned more or less with the No. 1 jaw of the chuck, just to make it easier to see where I am. But first you must rough machine the eccentric blanks and bore the hole. You must also pay particular attention to which way round the blanks are to be fitted. It doesn't matter whether the boss is outwards or inwards when attached to the fixture as far as use is concerned, but it does matter that the boss be on the *correct* side or you may find you have machined an angle of retard! If the blank is a casting, of course, the position of the cast lobes will be a guide; otherwise you must study the drawing carefully.

Set the blank on the mandrel and get the correct throw running more or less concentric. Tighten the nut and mark a line (not permanently yet) on the face, dead in line with the '120' line. Slacken the nut and rotate the eccentric until this line aligns with the 'O' on the fixture. Check that this lobe of the eccentric is reasonably concentric. If it is not, then you must adjust to reach a compromise then return to the original position, tighten the nut properly, and mark the setting line on the face of the blank permanently. Machine that lobe of the eccentric then rotate as before until the line on the blank's face aligns with 'O' and machine the back lobe. I won't go into details of the machining, as this depends so much both on the type of eccentric and on your own machining practice. Fig. 2.30 shows the job in progress on my own lathe. One point – it will, of course, be fatal if the blank slips on the mandrel! So, in addition to the locknut you may wish to use a spot of *Evo-Stik* or similar glue in

Fig. 2.30 Machining a twin eccentric.

the bore as well. I have never had one slip but there is always a first time.

A collet converting fixture

When referring to 8mm collets very few writers identify them by make, assuming, no doubt, that all are the same. This is far from being the case. They differ in draw bar thread, in keyway size

and, a few, in the nose angle, though these are not often met with. The table on page 79 shows the differences. Over the years I have collected quite a range of these collets as well as arbors for holding little cutters, drill chucks and the like. Some came with my little Boley & Leinen watchmaker's lathe, others with the Wolff-Jahn miniature milling machine, and others again from odd purchases at sales in case they came in useful. Most of these, I found, could be used in the Boley, but very few would fit all of the machines and accessories in the shop. I shall refer to the draw bar fitting later (page 77). This is not a difficult job to deal with, but the difference in the keyways did present problems. Apart from anything else there was the risk of damaging the key in the machine spindle if an undersize keyway was drawn in.

The fixture shown in Fig. 2.31 was designed to bring all keyways to the same dimensions. It is not all that

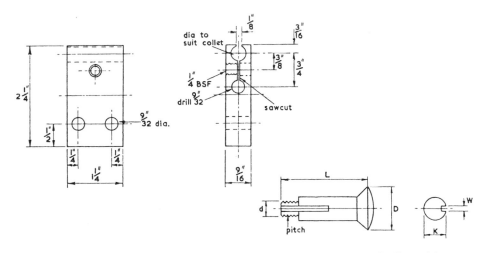

Fig. 2.31   Left: collet keyway milling fixture. Below: critical dimensions of collets of the same diameter.

important if a keyway is *over* size in this application as its sole function is to prevent rotation of the collet when tightening the draw bar. The dimensions of the block are not critical. I used $\frac{9}{16}$ in. thick because a piece that size was in the bitbox but $\frac{1}{2}$ in. would have served. After squaring off all sides it was marked out for the holes and then set up in the four-jaw chuck with the centre for the 8mm hole running true. The $\frac{9}{16}$ in. dimension helped, as it just fitted the groove for the chuck jaw. It is important that this hole be square to the front face of the block when in the chuck, and also that it be parallel to the $1\frac{1}{4}$ in. side. The hole was drilled and then bored to be a sweet fit on a solid arbor. This was used in preference to a collet

as the latter, being split, is not uniform in diameter when unloaded. An 8mm reamer would have been a help and if boring is the only way then care must be taken to get a good finish. After drilling and tapping the holes the block was split as seen in the drawing, the top of this split being opened out with a warding file to clear the cutter.

This was a 2.0mm Woodruffe keyway cutter, this size corresponding to the widest keyway (the Boley). The vertical slide was set up with more than usual care to get it square across the lathe (Fig. 2.32). The collet was set in the fixture with the shoulder behind the taper bedding against the end of the block. After setting the collet keyway in line with the slit (by eye) the saddle was

Fig. 2.32   The fixture in use.

brought forward until the cutter also aligned with the collet keyway and then locked. Trial cuts were made on a dud arbor, cutting starting from the head end of the existing keyway (so tending to drive the collet further into the fixture) and running right through the thread. Some collets had flats here so that no cutting was needed, but most had the keyway full length. Once the depth was correct (that for the Wolff-Jahn, the deepest of my collection) the vertical slide index was noted, and cutting started on the range of arbors. After the first few the average time taken was about 48 seconds. As the fixture had been set with its edge overlapping the face of the vertical slide a little it was possible to machine arbors carrying chucks without removing the latter – a great saving in time. Each collet of unknown provenance (some had no markings at all) was checked for hardness first, and two, which seemed to be dead hard, were left as they were. Any which seemed tougher than the others were left until last. *Macron B* cutting oil was used, and the cutting speed employed was about 25% less than that given in the *Model Engineer's Handbook* (published by Nexus Special Interests) for silver steel. The only problem met with was the mush of small chippings that got into the socket of the fixture, and a cleaning rod – just a bit of rag held in a twist of 16 gauge wire – had to be made to cope with this.

De-burring took some time, and was done on the cylindrical part with a fine India stone. Burrs on the threads did present some difficulty in a few cases. Those for which I had a die were easily dealt with but for others I filed the side faces of the slot with a No. 4 cut Swiss file first and then used a bit of softwood and some superfine emery powder in a little oil on the threads. The wood soon took up the thread profile. The collets were all washed in clean paraffin and then re-oiled. As remarked elsewhere (p. 77) there was little trouble with the draw bars. Two of the cutter arbors and three collets were rather stiff in the Boley drawbar, and these were eased with the die which came with the machine. I suspect that these ones had been made to fit the IME lathe, which has a thread about / thou larger in diameter than most. They had a Crawford number which I could not identify.

The fixture took about an hour to make, and the machining time was little more, but the result is that I can now use any collet or arbor on any machine, and this has vastly increased the range of sizes available. To have bought new collets of these sizes might have cost hundreds of pounds today. Well worth the time spent!

Round and About the Lathe

The lathe is the mainstay of our workshops, and rightly so, for it is the most versatile machine that has ever been invented – though invention is probably the wrong word for it has been the subject of development over several thousand years. The number of accessories and attachments devised over the years is legion; one manufacturer's price list contains one page covering the machine itself and seven more are needed to list the ancillaries. Anyone who has seen the two books on turning by Holtzapffel will know that the true ratio is nearer 700 to 1 than 7 to 1. However, it would be true to say that many of the artefacts made on a Holtzapffel would seldom find a place in a model engineer's workshop!

The following pages are not all concerned with 'things'; some deal with methods of using things rather than their design or construction. This is quite natural, for turning is a craft in which skills will develop over many years. The *way* in which tools are used is as important as the tools themselves. I have included detailed instructions on the manufacture of some of the devices. Some may think this unnecessary – for them a drawing is sufficient – but

not all model engineers have years of experience behind them. In addition, the machining instructions for e.g. the indexing knob of the screwcutting depth stop may have other applications as well.

It has proved unusually difficult to arrange the items in this section in any logical order, and there is inevitably some overlap of matter. Few model engineers' lathes experience a production run and if my experience is anything to go by, the making of models is often interrupted by essential repairs to domestic equipment or even the neighbour's hay-baler. You never know what may crop up in the machine shop so the arrangement of the subsequent pages follows the same pattern!

The ball centre
We are so accustomed to the standard 60° centre and the Slocombe drill to match that it comes as a surprise when some other form appears. However, it must be remembered that our friend the centre punch started life as just that – a punch for marking the centres in bars to be turned. The turner would take great care to ensure that his punch matched the angle of his machine's

centre. Few were at 60°, and I have some which measure 90°, one at 75°, and one which is markedly curved on the taper – all genuine and as the lathe-maker provided. The centre thus punched was often deepened or enlarged by using the square centre which acted as a negative rake countersink bit, and Myfords for one still list them. Our surprise is not, therefore, justified and we should expect to meet the odd one out especially when repairing old machinery or, as I seem to do quite often, when restoring an old lathe. There are many ways and means of dealing with these, but I find the use of a *ball centre* the most handy and the least trouble. This was first devised when an ancient tower clock was being overhauled, where the centres in some of the arbors (some over an inch in diameter) were found to have spherical centre holes.

This is how I dealt with this one. A hard steel ball was interposed between the work and the normal female centre. The diameter is not critical as long as the ball bears on the flanks of the taper where there is one. However, it must be a *hard* steel ball, from an old ball-bearing, and not the rustless balls supplied for use in valves etc. These will serve for light work but they wear rapidly. Worn balls are perfectly satisfactory provided they are not chipped or too badly pitted. The method has since been used on a variety of repair jobs with perfect results provided, of course, the original centre hasn't been burred over or damaged. It saves the chore of setting up the fixed steady and re-cutting a new centre hole.

Setting over the tailstock
This is always a nuisance. The actual setting over is little trouble, but I must confess that I detest having to do so after I have taken so much care to adjust it to turn truly parallel. Furthermore, the less my parallel test mandrel is used the better, for although it is precision ground it is not hardened. For many years I had in mind to design a calibrated attachment to fit to the tail-stock poppet but never found the time, until one day it occurred to me that I had one all the time in the shape of my boring head fitted to a No. 2 MT shank *and* calibrated in thousandths of an inch! See Fig. 3.1.

This accepts $\frac{3}{8}$ in. diameter round-shank cutters and all that was necessary was to make a 60° point and harden it. In the event I made it slightly more than 60 on the grounds that when in use, set over towards the front of the lathe, it would bed better that way. The next job I had to do required the setover to be backwards, so this was a waste of time and 60° would have served both ways round. The point was made of silver steel (known as drill stock in the USA and Canada), machined by setting the boring head in the headstock taper with the index at zero so that the centre-point went in the same way each time I filed a small flat. Here I should mention that many failures to reach adequate hardness are due to the hardening temperature not being held long enough. The transition from one state to the other when hardening takes time, and it is not only necessary to heat slowly but also to *hold* the final temperature for about five minutes for each one-eighth of an inch of thickness (or diameter) if the work is to be hard right through.

To prevent scaling I use an anti-scale paint obtained from the makers of ceramic enamels, but a pasty mixture of powdered chalk and water (or meths)

Fig. 3.1 Using a boring head as an offset centre.

will serve, the only snag being that you cannot then judge the temperature so easily. Otherwise you will just have to tolerate the scaling and remove it with emery afterwards.

This device has been used for taper turning now for 25 years or more and it has proved perfectly satisfactory. To be on the safe side I don't rely on the readings of the micrometer index, but check with a dial indicator; there is a slight difference sometimes. However, although I *have* used it for turning fish-bellied rods (see page 53) I would not apply it to any which were above about $\frac{3}{16}$ in. diameter, as the side forces would then be too great.

The spring centre
Sooner or later every writer of descriptive articles will tell you to 'support the

tail of the tap from the tailstock centre' when tapping a concentric hole in the lathe. This is all very well, but as soon as a slight bit of thread has been cut the tailstock ceases to perform its office while the regulation reverse rotation of the tap to clear chippings is hazardous.

It is possible to have one hand on the wrench and the other on the tailstock handwheel but this is difficult. This little gadget is designed to overcome these problems. As you can see (Fig. 3.2) it is no more than a spring centre and is easily made. The spring holds the centre-point engaged with the tap for perhaps $\frac{1}{4}$ in. or so of travel, after which the tailstock barrel can be re-adjusted. I know that some taps don't have centres at the tail and I use a Jacobs chuck in such cases, however the wrench has a habit of slipping. If

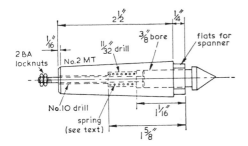

Fig. 3.2 *Arrangement of the spring centre.*

you have a Quorn, I suggest you set up and grind a *male* point on the tap tails and use the female centre shown in the inset sketch. I find that most of my taps above 2 BA do in fact have centres.

Make the centre first. You will see (Fig. 3.3) that the shape is different in the arrangement from that shown on the detail – entirely a matter of choice. Chuck a piece of $\frac{1}{2}$ in. dia mild steel and face the end. Centre, and drill and tap the 2 BA hole for the retaining stud, then turn the sliding barrel to 0.375 in. dia by 1 in. long, getting a really good finish and making sure it is parallel. Make a little recess at the shoulder. Reverse in the chuck, using paper to get it running true, take a skim over the $\frac{1}{2}$ in. dia and form the shape of the end. Get a reasonable finish but we shall true it up *in situ* later. Now make the retaining

stud (this is to prevent the centre from falling out and getting lost in the swarf and to serve in the final stage of manufacture) using the tailstock die-holder on the threads.

Making the socket

The socket is made from a soft Jacobs drill-chuck arbor, preferably one with the Jacobs No. 6 chuck taper (chuck Nos 6A and 34) or larger. You will notice that the tang has been cut off – this is not essential but it makes matters easier. You can't bump the centre out anyway (hence the flats on the head end). Set the arbor in the headstock socket and machine the chuck end parallel, say $\frac{5}{8}$ in. dia as far as the beginning of the Morse taper. Now drill the No. 10 clearance hole right through. Part off the surplus of the chuck taper, facing the end to a good finish. Drill $\frac{11}{32}$ in. dia × $1\frac{5}{8}$ in. deep, and then bore to a smooth fit to the centre you have already made to $1\frac{1}{16}$ in. deep. Very slightly bevel the edges of the hole. Remove from the lathe and file two flats to fit any convenient spanner.

Fit the retaining screw to the centre and, using washers if necessary, draw the centre hard up against the shoulder in the arbor; then refit in the headstock. Take a final skim over the point – you now have this dead true. Whether you harden the end or not is up to you. It

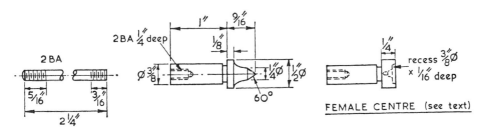

Fig. 3.3 *Details of the spring centre.*

51

won't get much wear, but it may get knocked so I recommend you case-harden it using *Kasenit* or a similar compound. Wrap a length of iron wire round the parallel shank and clart it up with powdered chalk made into a paste with oil or water, or use a proprietary brand of anti-scale paint to stop the shank from being heat damaged. When hardened – and I recommend a double dose of *Kasenit* – clean up and chuck the part to polish the working surfaces.

The spring is made from 20 or 22 swg steel wire with 11 turns wound on a $\frac{1}{4}$ in. dia mandrel, the ends ground square. However, you may care to experiment a bit. If most of your work is done with taps below $\frac{1}{4}$ in. then a lighter spring will prevent excess pressure when starting small taps, but if you use many taps over $\frac{3}{8}$ in. you may find the spring pressure not sufficient to prevent the centre from sliding back when turning the wrench. The solution is, of course, to have two (or more) springs. Incidentally, although I mentioned a $\frac{1}{4}$ in. mandrel you may have to reduce this a little with the heavier gauges of wire to get the spring to enter the hole.

The works should be assembled with the two locknuts adjusted to allow the centre to slide outwards between $\frac{1}{4}$ in. and $\frac{3}{8}$ in.; no more than that or the side thrust will be excessive. In use, the tap is started either with maximum spring pressure or by screwing the tailstock forward until the shoulder butts on the arbor. Once started, the tailstock barrel should be retracted about $\frac{1}{16}$ in. and moved forwards as required – about every $\frac{1}{4}$ in. of tap penetration. The tap can be worked back and forth quite freely, and the tailstock barrel can be locked if desired.

To make the female centre, finish this with the boss faced, but not centred; leave this until last and drill a little centre with a Slocombe, with the whole assembled in the arbor, as you did when finishing the male centre. There is less need to harden the female part, as the hole can't get knocked!

The device has other uses, as no doubt you will find. It is a handy 'holder-up' – for example if you want to lay a rule across a job on the faceplate. It can hold two pieces together while *Araldite*, *Loctite* or chucking cement is curing and I find it more useful than a plain centre when spinning work in the lathe to check for truth under a DTI. However, please don't try to use it as a substitute for a running centre – it just isn't man enough for that sort of work!

Drilling an arbor for a draw bar

The Morse taper socket of the lathe is a wedging taper and any arbor properly fitted should grip sufficiently to prevent any rotation. However, in most milling, and some drilling, operations there is a force which can tend to pull the taper from its socket. This is why milling cutters are best held in the three-jaw chuck rather than a drill chuck. I hold most drills that way too, but there are times when the ordinary drill chuck must be used. A similar consideration applies to arbors carrying boring heads and the like. The problem is that chuck arbors drilled and tapped for draw bars are somewhat expensive; most are fitted with tangs and it isn't easy to see how to drill and tap the 'wrong' end of the arbor. You can, of course, set it the right way round in the chuck and drill right through, but this means buying or making an extra long drill.

I suggest you try it this way. Set the arbor between centres with the tang at the tailstock end. Turn a short parallel portion just above the tang, about $\frac{3}{8}$ in. long. The diameter doesn't matter but the finish should be good. In carrying out this part of the work, take care that the lathe carrier does not mark the chuck taper at the other end. While still between centres, set up the fixed steady and adjust it to this short parallel part. Then, after sliding back the tailstock, slide the steady along the bed out of the way for the present. Cut off the tang.

Attach the drill chuck to the arbor and grip in the three-jaw chuck by the cylindrical part of the fixed body, somewhere between the keyholes. Bring the fixed steady up to the arbor and secure it. You can then safely drill the end of the arbor and tap it say $\frac{5}{16}$ in. BSF. Most taps this size have a centre in the head, so you can guide it with the tailstock centre to ensure that it runs true (but see page 50). Finally, machine the end of the arbor with a slight radius at the edges. It will help if a 60° centre is offered to the tapped hole; this will guide the end of the draw bar while fitting.

The draw bar itself need be no more than a piece of studding, but it will help if the rear end is supported in a bush in the hollow mandrel. Brass is best as this won't damage the mandrel. My own draw bars are 'permanent' (see page 121) and have a cap nut secured at the end, and the same draw bar fits several boring heads, chuck arbors and the holder for the little Clarkson throwaway slot drills I use. On the old ML7, which had a hollow tailstock, I used a similar arbor for the tailstock drill chuck and so never had any problems with scoring. In passing I would mention that the most usual cause of rotating drill chucks is the use of drills with full 'rake' when drilling brass, particularly when breaking through. This material does 'grab' (hence the use of zero rake when turning) and if you haven't got straight flute (or slow helix) drills especially for brass it pays to stone the acute cutting angle a little to reduce the rake.

Turning fish-bellied rods

Models of many early engines require parts to be turned to a fish-bellied curve, examples being parallel motion links, connecting rods and parts of valve gear. These can be machined using the handrest, but if in steel this requires some experience – particularly if the part is slender. The use of a file is to be deprecated and, in any case, will need hours of work to remove the rough marks and it is very difficult to obtain a symmetrical shape by operating cross- and top-slides simultaneously. The following method has been applied successfully to all materials, including stainless steel, provided the work is not too stiff. It relies on the fact that a cantilever, loaded at the end, deflects very nearly to the arc of a circle.

Fig. 3.4 is an example. The procedure is first to cut off a piece of stock rather longer than is needed – not less than two diameters each end – and after facing both ends to centre fairly deeply with a Slocombe drill. Set up between centres and turn the full length to about 0.005 inch above the largest diameter. With a sharp-pointed tool cut a tiny groove to mark the position of the shoulders A and B. With a felt-tipped pen mark the position of maximum profile, C, which is not necessarily in the middle of the length.

Refit the three-jaw chuck and set the tailstock over *towards you* by an amount equal to half the difference in diameters at C and A or B. Grip the stock in the three-jaw with the section at C just inside the jaw faces. Fill the centre-hole

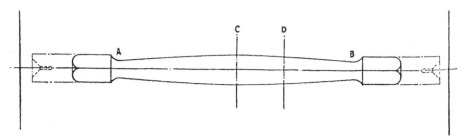

Fig. 3.4

with a mixture of tallow and either graphited or 'Moly' based oil and engage the tailstock centre, with the poppet projecting no more than is absolutely necessary. You can angle the topslide if necessary and use the leadscrew handwheel (or, better, power feed) for traversing. The length BC will now be curved towards the tool and as the latter will travel in a straight line the work must adopt the curved profile shown. Note that the tailstock centre is very heavily loaded and must be kept well lubricated. I do not recommend the use of a running centre as this will almost certainly get in the way, and in any case not all such are designed to take *side* loading of any great magnitude.

Machining procedure depends a little on the stiffness of the work, but all tools must be honed to a fine finish and, with steel, ample cutting oil applied. When machining the connecting rods for 'Rainhill' I plunged down with a round-nose tool almost to diameter at the base of the curve (A and B) and then 'shifted metal' with a knife tool; this puts all cutting forces in an axial direction away from the loaded tailstock centre. The final cut was made with a round-nose to finish the end radii and to get a good surface on the rest. I always use power feed when I can, and this gives the desired machine finish

looked for by the judges at exhibitions!

The work is now removed to machine the other end. If BC is shorter than the chuck jaws it can simply be reversed and the operation repeated. If the diameters at A and B are not the same, then the tailstock must be readjusted to suit. There is little difficulty if the ends ought to be smaller than the diameter at C, as these can be turned down afterwards, but if C is the smaller then a collet must be made to fit over the workpiece at C, split, of course. The important point here is that the edges of the split must be smooth and well rounded or the work will be badly marked. Finally, a problem may arise if the length AC or BC is greater than that of the chuck jaws. In this case a holding jig must be made. This is no more than a piece of steel about twice the diameter of the job, drilled and preferably reamed to fit the work, and slit longitudinally. Again, it may be necessary to fit a ring inside this collet if the diameter at C is less than at the ends.

It remains to deal with the short parallel portion which is left at the centre of the length. Often this may be barely noticeable, and in many cases I make a feature of it as shown in Fig. 3.5; you cannot do this, of course, if working to a known prototype, but many 19th century engines were, in fact,

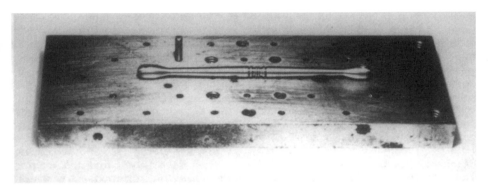

Fig. 3.5 Added decoration disguises the discontinuity at the centre of the link.

made that way. The decorated pattern can be turned between centres in the ordinary way. However, if an exact blend is needed the following procedure is very effective. Set a very sharp round nose tool (not too great a radius) in the slide rest and offer it to the work between centres until it *just* shaves the diameter at C. Traverse back to the point where the previously machined curve just ends without adjusting the cross-slide and apply thumb pressure on the back of the work until the tool again just shaves the work. (You may have to use a piece of wood if your thumb gets too hot!) Apply longitudinal feed either under power or with the top-slide and as the tool traverses gradually release the thumb pressure. Now repeat for the other side of the centre section, traversing in the opposite direction. A little practice on a piece of scrap will help judge the amount of pressure required, but the procedure is surprisingly easy. So much so that in some cases I have done the whole job this way, dispensing with the tailstock altogether. The one thing to watch out for is that you don't get a piece of swarf between work and finger, hence the suggestion that you use a piece of wood!

The finishing of the ends follows normal practice, the unwanted surplus being parted off and, if necessary, a tiny centre being left in to complete the operation. (Don't forget to reset the tailstock, though! See page 49.) The method is very successful, the only proviso being that the part be slender enough for the offset tailstock to be able to bend it. You could not expect to be able to machine a stiff column this way.

There are cases where a curved taper runs the full length. If that is so, then you will need a length of stock greater than that of the finished part by an amount about equal to the axial length of the chuck jaws. It is better to use the four-jaw in this case. This can be parted off later. It is essential that the chuck gets a good grip of the work or it will not take up the desired shape. In cases where there is a straight taper at the larger end, only the top two-thirds or more being curved, proceed as normal, but terminate the matching at the required point – say D in Fig. 3.4. The taper should be machined full length first.

Where the bases of the part are square the same procedure is used with the four-jaw; obviously this needs more

55

care as the chuck must be reset each time. I am fortunate in having a self-centring four-jaw, bought secondhand many years ago. It is worth its weight in gold; not particularly accurate, but it does save an enormous amount of setting up in the many situations where the cylindrical part of the workpiece need be no closer than about 3 thou coaxial with the square ends.

Crankshaft machining aids

There are three main problems when machining crankshafts: (a) the long overhang of the tool, especially when the throw is large, (b) the accurate setting-out of the throw-pieces, and (c) preventing deflection of what is a very long and slender workpiece – especially if it is a multi-throw crank. Some contribution to the solution of (a) is given on pages 93 and 95. To meet (b) I always use the co-ordinate method, drilling the necessary centre-holes, and even boring, either on the vertical slide or on the milling machine. The following notes may help with the rigidity problem, (c).

If the crank throw is solid then there will be no difficulty in the initial machining of the journals – the shaft can be set between centres with no risk of bowing. However, if, as is good practice, a final skimming cut is taken over the journals after all other machining is done (for whether a forging, or chewed from a solid piece of metal, there is likely to be

some movement as internal stresses are relieved) then the lack of stiffness at the web or webs will be felt. A packing piece should be set between the webs, however, the system can be stiffened up still more if clamping pieces are used as well, and these have the advantage that the cross-bolt prevents the packing from failing out. See Fig. 3.6. Note that the clamps bear on three points, one on one web and two on the other. This ensures that the clamps themselves do not impose any distortion on the shaft.

When machining the crankpin, packing must be set between the webs and the throw-pieces to take the thrust from the tailstock and also to stiffen up the assembly. Fig. 3.7 shows one solution and Fig. 3.8 another; the first for a slender three-throw shaft and the second for a stiff single crank model of some size. I always provide a centre-hole in the throw-piece for the end of the Fig. 3.7 type jacking screw to sit in. This is located without any attempt at real precision; a small drill is taken right through the throw-piece from the centre-hole drilled for the lathe centre, and (if need be) opened out with a Slocombe on the back.

Additional stiffness can be achieved (and is strongly recommended when dealing with a multi-throw shaft) if clamps as used when machining the journals are fitted to all webs except

WEB PACKING.   CLAMP PLATE

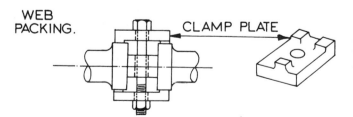

Fig. 3.6 Use of clamps on crankweb to stiffen the shaft while machining.

Fig. 3.7 Purpose-made jacking screws to take tailstock thrust.

that which is being machined. Mention is made elsewhere of the tool shape used for crankpin machining when dealing with tangential tools (page 95 and Fig. 3.44) but a word about measuring the pin may be helpful. It is often

Fig. 3.8 Toolmaker's jacks in use on a 1-inch single-throw crank.

not possible to get a micrometer onto the pin, and even a vernier caliper may be difficult. Even when the anvil of a micrometer can be applied it may be difficult to be sure that the pin is parallel. In such cases I start by machining a short stub of steel to two diameters, one the desired size of the pin, and the other 0.004 inch more in diameter. A piece of $\frac{1}{16}$ inch gauge stock (or even mild steel or brass if it is only going to be used the once) is then filed with two gaps each a sweet slide fit to one of these diameters. This can then be used as a gauge when turning. If the large (oversize) gap 'goes' then I know that I am within less than 0.004 inch of drawing size, and can proceed accordingly. If it 'goes' too easily, and the other gap doesn't, one has to be careful! I then use feelers in the gauge to check the size before taking off any more.

Every shaft is different, and it would not be right in a book like this one to go through the various methods. But one point, already hinted at, is worth emphasising. A fair bit of metal is removed when making a crankshaft, and in its very nature there will be locked up stresses in the rough material, whatever form it takes. It pays to rough machine all over, and especially clean up the webs, first, and then leave the job for some days before proceeding to finish machining. I would never cut a crankshaft blank from bright drawn mild steel without stress relieving at about 350°C or at the hottest, barely visible red, for about half an hour and would still allow a gap of some days between roughing out and final machining.

A centre-height gauge and scriber

A number of devices for setting lathe tools to centre-height have been described in the past, but I make no excuse

Fig. 3.9 Combined centre-height gauge and scriber for use on the lathe.

for presenting yet another, for it offers several advantages. As will be seen from the photograph Fig. 3.9 it is double-sided, so that it can be applied equally well to tools in the front or rear toolpost. Secondly, the hard HSS points of the gauge are available as an accurate and preset scribing block for marking out work in the lathe. There is, for example, no difficulty in setting out rings of holes using the jaws of the three- or four-jaw chuck as a dividing agent (see page 133), and really accurate marking out can be done by using the device in conjunction with the vertical slide; the thing then becomes a sort of vernier height gauge, but with the work rather than the scribing point being moved.

The dimensions are given in Fig. 3.10,

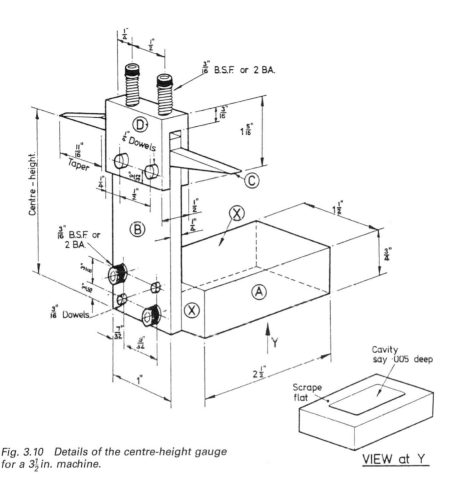

Fig. 3.10   Details of the centre-height gauge for a $3\frac{1}{2}$ in. machine.

VIEW at Y

but none of these are vital and will depend on the size of HSS toolbits which are available. Those I used are made from a single piece of $\frac{1}{4}$ in. square material that was nicked on the grinder and snapped in two. The ends are then squared very carefully on the grinder, and the long sloping bevel is ground, leaving a 'witness' of the squared end a few thousandths of an inch wide. The final sharpening of the end is done with

a very fine oilstone and much care, so that the 'point' is both sharp, and square across the end. The burr underneath must also be removed with the same fine stone and equal care.

The base part (A) started life as a piece from the junk-box and is steel, though a bit of cast iron would be better. But pick a piece which has not been knocked about too much, to save work later. The first step is to make all faces as square as possible with a file (or with a milling cutter, of course) and then pick three of these faces, marked (X) on the drawing, as being the most square to each other for further work. Note that the actual dimensions are unimportant. Now start with the underside of the base, and with a suitable file excavate a cavity a few thou deep in the centre; leave a 'frame' about $\frac{3}{8}$ in. or $\frac{1}{2}$ in. wide all round untouched. This will save work, for the next step is to bed this absolutely flat using scraper and marking blue on your surface plate, or, if you haven't got one, on a piece of plate glass, or even the lathe bed for want of anything better. If the latter, use the part at the tailstock end which is least worn.

Now file the two upright faces – those shown $1\frac{1}{2}$ in. and $2\frac{1}{2}$ in. long – and, if necessary, scrape them until they are square to each other and to the scraped base. Note: it is most important that the end face be squared to the base; next most important that this face be square to the long face; and least important that the long face be square to the base. This having been done you may now 'prettify' the block – Duplex would probably frost it with a scraper, but I did no more than clean it up with emery.

Now for the upright, part (B). This must be of material the same width as the toolbit, and may be bright drawn steel, brass, or what you will. I used a bit of ground gauge-stock that was handy.

Clean it up all over, but try to avoid reducing the thickness to below that of the toolbit. Square off one end as true as may be. Now mark out for length. This must be equal to the centre-height of the machine, **less** the thickness of the toolbit **less** again say $\frac{1}{32}$ in. – this last

subtraction to ensure that the bottom of the upright is clear of the lathe bed in service. The top end must now be *accurately* squared across the thickness (otherwise the scriber blade will lie slantways, with consequent loss of accuracy) and both flat and square in the longer direction; the flatness is more important than dead squareness, but make it as near as you can. Drill the two holes for the bottom fixing screws, with rather more clearance than normal, but don't drill for the pegs yet.

The stirrup piece (D) is the most difficult part. If you have a cutter which is the right width and which will reach deep enough, this will ease the job, otherwise you will have to attack the piece with drills, saw and file and make the slot. This should be a reasonable fit on the toolbits and the upright, but it does not matter if it is a trifle easy; it will not hurt to 'close' it a bit in the vice to make it a slide fit on the upright. Clean it up so that it looks reasonable, and drill and tap the two screw-holes. Take care in this job to get the holes square to the work, and tap with care to get a good fitting thread. If the two screws to be used have not already a radiused end, spin them in the lathe and put a smooth end on.

You may now drill for the two dowels which hold D and B together. Put a piece of packing in the slot so that when the job is assembled there will be $\frac{3}{32}$ in. to $\frac{1}{8}$ in. gap between the upper scriber bit and the top of the slot in service. Clamp all up to part B and drill the holes right through, finally reaming $\frac{1}{4}$ in. for the dowels. These are made from silver steel. They need only be a good push fit – they are not **driven** in; this is unnecessary. Now drill and tap the holes in the base for the screws attaching the upright. Set the latter square on the base, and clamp up so that the lower end of the upright is about $\frac{1}{32}$ in. upwards – set the block on the lathe bed and the upright with a 30 thou feeler underneath – and clamp securely.

Spot through with the clearance drill already used, to mark the base, and then use the tapping drill and finally tap the holes. Make the holes about $\frac{1}{4}$ in. deeper than the screws are long. Assemble the two parts, using the square, and gently pinch up the screws. Do not fully tighten them.

Now to set the height; if you have a vernier height gauge, see later. Measure the toolbit – it won't be exactly to dimension, but probably a few thou down on the nominal thickness. Set a piece of waste material in the three-jaw – or between centres, if you prefer – and machine a cylinder which is exactly *twice* the dimension measured from the toolbit and a little over 1 in. long (or width you have made the upright). It is important that this test cylinder be parallel and, of course, the accuracy of the diameter will control the accuracy of the gauge. Now comes the tedious part, but if you take care here it is worth it. You must adjust the upright so that it *just* touches the underside of the cylinder full length. This means much wiggling the thing about, and tapping with a hammer this way and that until you get it right, and it may prove stubborn! However, if you check with feelers and, say, find that it is parallel but .005 in. low, then set the assembly on a flat block, fit feeler strips below the upright until they just slide in, add another .005 in., and give a smart tap to the base. You must keep on trying until you get it right, and then tighten up the base screws. *Warning*: do this a bit at a time, and check the accuracy afterwards, as

the tightening of the screws can cause the upright to move. Once all is correct, and the screws tight, drill and ream for the two dowels. These should be a good fit, but check they don't shift the upright.

If you have a vernier height gauge, of course, the job can be done with that, and in this case – provided it has a reversible index point – you can set the gauge after assembling the two toolbits into the stirrup, and set to the points. Note, however, that in using this method you may be taking a bit of a risk, especially if you have an older lathe. The centre-height may not be as it once was, and in any case even new lathes are made to a tolerance. The turned bobbin method allows for this.

Final assembly calls for little comment, but once done you should make a further check. Set up a fairly wide piece of material in the lathe – a chuck backplate held in the three-jaw will do – and scribe a line right across with the centre-height gauge. Turn the mandrel through 180° and adjust it until the scriber point just coincides with one end of this line. Transfer the scriber to the other end of this line, and it should still coincide. If it doesn't, then the gauge is in error by half the difference between the line and the scriber point.

If an error does show, repeat the operation on another part of the back-plate using the second scriber point. If the error is in the same direction in both cases, then the upright as a whole is too high or low (you can tell which by inspection of the lines). If one shows 'up' and the other 'down' (or correct) then the upright B is slantendicular, and can be tapped sideways to correct. Note that this method of checking shows the error doubled; if the difference is only just discernible it may well be no more than .001 in., but if the scriber points have been properly sharpened, an error of .00025 in. can clearly be seen – this would result in a difference on the far side of the plate of .0005 in., and such is easily distinguished.

Since I made my gauge the price of HSS toolbits has rocketed, and you may care to consider the use of carbon (silver) steel or gauge plate. This is quite in order and, in fact, carbon steel is slightly harder than HSS, but it does mean more work though less wear on the grinding wheel. The taper nose can be filed to shape before hardening, but leave the end blunt; a face about $\frac{1}{16}$ in. wide. There is no need to harden the whole issue, just the last half-inch taper part. Heat to 780°C for 10 to 15 minutes, to give time for the heat to soak right through and for the metal to change its character, then quench, point down, in water. Don't swirl it about – a gentle up and down movement is better, but don't withdraw until you are sure the metal is cold. Clean off the scale and then temper by boiling in a deep-frying pan for about 15 minutes.

There is a slight risk that the piece may bend, so the bottom of one and the top *and* bottom of the other part must be ground flat. I suggest you do this with grinding paste on a flat piece of glass or, better, a lapping block if you have one. Check the thickness of the bottom arm with your micrometer and ensure that it is parallel. Fortunately the shank part will be relatively soft, and a *sharp* scraper will enable you to correct any bulges without difficulty. Finally, form the sharp scribing points with a fine India oilstone. Take care to keep the corners sharp, as these act as the scribing edges.

Note that it will not pay to have the points projecting more than that shown in the photograph; the longer the points the more accurate the rest of the workmanship must be. On the other hand, if they are much shorter you may find there are occasions when they won't reach when marking off.

In use, there is only one precaution necessary – apart from making sure that there is no dirt beneath the base – that is to take care not to ram the scriber points hard on to a tool edge. This will damage both. Finally, you will see that there is no reason at all why you cannot remove the toolbits and replace them, perhaps with longer ones for special jobs if you wish. The accuracy depends on the setting of the upright member, and provided the replacement bits are of the same size, this will be retained.

A micrometer screwcutting depth stop
Before describing the device, a word about the screwcutting operation may be desirable. Many authorities recommend that the topslide be set over by half the thread angle, the idea being that the tool then cuts on one flank only. This, it is claimed, avoids the problems of the congestion of chips at the point of the tool, enables rake to be applied to the active side of the tool, and produces a better finish on that flank as a result. Frankly, I doubt if either consideration is of much importance to model engineers as the majority of our screwcut threads are very shallow at 32 or 40 tpi.

However, to deal with the last two points first. A little reflection will show that, with right-hand threads, the left-hand flank of the thread cannot touch its mating thread – the nut or whatever. Once the threaded pair is tightened the unavoidable clearance is all on the left-hand flank – that on which this method claims to give a better finish. Second, there is already a slight rake on this left-hand side (about 3° with most threads) and this appears as a *negative* rake on the thrust-bearing flank of the thread. Any increase in the cutting rake produces an increase in negative rake on the side of the thread that matters, so that even if a final, shaving, cut is made the finish on the side of the thread that carries the load will be inferior. If any rake is needed it is on the opposite side to that suggested by those advocating this method.

So far as 'chip congestion' is concerned, if the nose of the tool is rounded (as it should be) then this will be cutting a curved chip anyway. But finish cuts when threading are very light, and even when cutting heavy threads there are other ways of overcoming the problem. The difficulties experienced by many turners are due more to lack of rigidity at the tool point (or of the workpiece) resulting from, as a rule, excessive tool overhang or too light a tool section.

The preferred method of screwcutting is to use a tool which has the top surface ground with a slope rising to the *right-hand* edge at about 3° – or the appropriate pitch angle if that is known. This means that both cutting faces have zero rake. If this slope is increased slightly there may be an advantage with certain metals; I use 5° for silver steel. The active flank of the thread then has a better finish, at the expense of that on its less important face. For deep threads (say below 14 tpi) take a fairly deep initial cut. Return the carriage and, at the *same depth*, repeat the cut with the topslide (which is parallel to the lathe centres) advanced towards the headstock by 2 or 3 thou. This widens the groove. Repeat again, but with the topslide adjusted again to take off a couple of thou on the opposite side. This 'gashes' the thread and not until the tool has advanced in depth by the amount of the initial cut will the tool engage fully again. This process is repeated, using smaller and smaller depths of cut as full depth is approached. Every two or three increases in depth the topslide is moved towards the headstock by about 0.001 in. – or the reverse, it doesn't matter much – until the final cut is reached. The tool is then allowed to 'shave' both flanks, and the increased rake on the right-hand side will produce the fine finish needed there to carry the load on the screw.

This method gives a better finish, and avoids the need to set over the topslide – and, more important, the fiddly business of resetting it to turn parallel for the next operation. As screwcutting is usually an intermediate operation in the making of a part this is not always easy, for it may mean unchucking the workpiece in order to fit a test mandrel. But it doesn't get over the main problem, that of remembering the dial setting on the cross-slide at each pass; even more so when, as the description below indicates, there are distractions! The first part deals with the ML7, and the second with the modifications I made when fitting one to the Super-7.

I recommend that you read the whole before starting on any work. As will be explained later, I redesigned the device after changing from the ML7 to the Super-7. It may well be that the later design would fit equally well to both, but as the ML7 had gone I had no means of checking. The Mark 1 (ML7) type cannot be fitted to a lathe with power cross-feed.

Those readers who followed the articles in *Model Engineer* describing my workshop will recall that it included a small bench for the daughter of the house, Georgina, then six years-old. Being young and enthusiastic – and female – her work was always accompanied by a running commentary on the behaviour of tools or workpiece, interspersed with requests for assistance. Always entertaining, often amusing, but occasionally most distracting, especially when trying to remember the last dial setting while screwcutting!

Now this problem never arose with my Fenn ornamental turning lathe (housed in the old workshop). In common with my Holtzappfel, and other similar machines of 100 years ago, the cut is put on by lever, with a screw stop which acts as a restrainer to the depth of cut, and a second similar screw with index on the head for use as a final depth stop. There is no alteration to the setting of the restraining screw when taking off the cut, and no feat of memory required. (Only one of calculation, for the thread on the Fenn was 22.18 tpi!) A little study showed that an hour or so's work would provide a similar facility on the Myford, and very effective it has proved.

Fig. 3.11 shows the arrangement. The slide-bar (A) is provided with a slotted hole and is held by the Allen screw and strong-back washer (2) (3) using the hole in the saddle to which the travelling steady is attached. This bar is adjusted to suit the diameter of the workpiece, and that shown will allow for cylinders up to $3\frac{7}{8}$ in. diameter. The bracket (4) is slipped in between the end of the cross-slide and the feedscrew bracket, and has a tapped boss at one end to carry the stop-screw (5). This has 40 tpi, and the head is engraved with 25 serrations engaging with the click spring (6). Thus one click

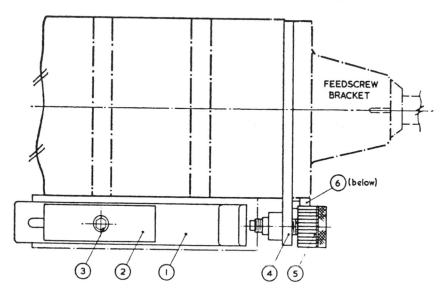

Fig. 3.11 Arrangement of the screwcutting depth stop.

(it can both be felt and heard) is one thou. The travel on this need be no more than the depth of the largest thread to be cut, but the drawing shows a travel of about 0.2 in. Just as easy to make, and one never knows – some local farmer *may* one day have a breakdown requiring the cutting of a $3\frac{1}{2}$ in. Whitworth screw!

The first step is to see that you have, or can obtain, a pair of Allen screws longer than those holding the feed-screw bracket by the thickness of the stop-screw bracket. (These screws are 2 BA by the way and **not** $\frac{3}{16}$ in. BSF; very close, and easily mistaken.) If not, you will have to make do with cheesehead screws while waiting for them. The bracket (Fig. 3.13) was made from a piece of $\frac{5}{32}$ in. × 1 in. bms flat – $\frac{3}{16}$ in. would be better as there is a very slight deflection if the cross-slide is wound in

fast, and one click will give a feed of about 1.2 thou. However, the device is not intended to be bumped in this way, rather treated as the micrometer which it is! (A preferred alternative is suggested in Fig. 3.16, as fitted later to my Super-7, see page 72.) File this down to $\frac{15}{16}$ in. wide, square the ends and clean up the faces. Mark out a centreline across the workpiece $1\frac{11}{16}$ in. from one end, and from this set out for two $\frac{7}{32}$ in. holes at $1\frac{15}{16}$ in. crs, $\frac{5}{16}$ in. from the top edge, to form the ends of the two slots. (Put a centre-pop at the end of the centreline to mark which **is** the top edge.) Mark out for the $\frac{11}{32}$ in. hole into which the screwed boss will be fixed.

Using the vice, with a bit of scrap steel beneath the work, first deepen the centre-pops with a Slocombe drill, and follow with a drill of the desired size. Follow with a 90-degree rosebit to

Fig. 3.12 The stop fitted to an ML7.

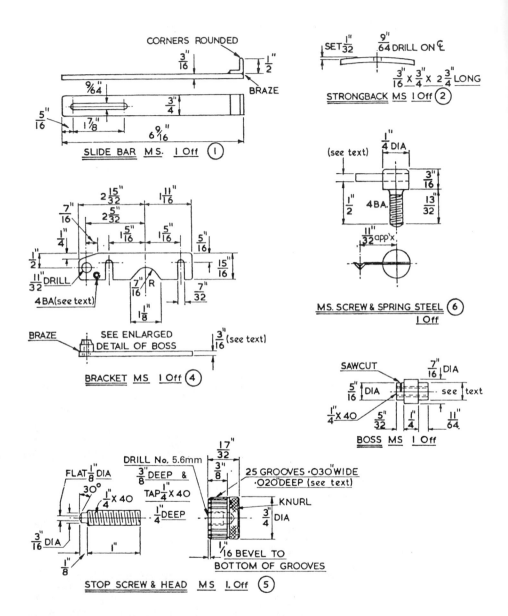

Fig. 3.13   Details of the depth stop.

remove burrs both sides. Put a decent countersink on the front of the $\frac{11}{32}$ in. hole, as this will make a better job of the brazing later. Now with a fine saw take out most of the metal from the slots, and put a cut up the centreline to guide the file when making the centre aperture. Clean out the slots with a warding file, making sure that the edges are at right-angles to the bottom face of the bracket, and remove the sharp edges.

With the piece upside down in the vice, attack the centre aperture (this is to clear the feedscrew nut) with a round file. Don't nibble at it, use a ten-inch bastard. If need be, start the groove with a three-square in the saw slot, especially if you are not well-blessed with sharp files. Once you have reached the depth of $\frac{7}{16}$ in., open out the sides with a flat file to the dimension given, and then finish off nicely and remove burrs. Incidentally, this business of filing always seems to be one of the bugbears of the model engineer – and not a few 'tradesmen' as well. To be able to file flat requires experience (and the right file; you can't file flat if the file is flat!) but even professionals at the job seem to make two simple mistakes. The first is referred to above – that of 'nibbling' with a small file. The second (and this applies to saws also) is to forget that a file is a cutting tool just like those in the lathe. There is a correct cutting speed and it is quite low – files are carbon steel. A 14 in. file will allow a stroke of about a foot. A cutting speed of 80 ft/min. means about 40 strokes per minute (you have to draw the file back, remember) or $1\frac{1}{2}$ seconds each. This sounds very slow work, but if the pressure is right the metal will pour off, and so will the sweat after about 10 minutes! A third point – always reserve your new files for brass, transferring them to the shelf for use on steel when they start feeling slippery on brass. After which homily, you should have no difficulty in shaping the end of the piece, rounding the sharp corners and giving a final polish, using emery if you must.

I have said nothing about the 4 BA tapped hole shown on the drawing. Don't tackle this until you have made the spring detent, as you may have to alter the dimensions given.

Make the boss next; a simple turning job. Chuck in the three-jaw, face the end, centre, and drill No. 2. Grip a $\frac{1}{4}$ in. × 40 taper tap in the drill chuck, disengage the mandrel drive, and using a delicate hold on the chuck start the thread in the hole; don't go right through. The object is to make sure that when you do tap the hole it is true with the axis of the boss. Starting the thread will guide the tap later. With a knife tool turn a shoulder that will just enter the hole in the bracket. It won't be exactly $\frac{11}{32}$ in. The boss so formed should be about $\frac{1}{64}$ in. less in length than the thickness of the plate. Reverse in the chuck, face off to length, and turn down the end to form the little boss shown on the drawing. Don't make the sawcut yet.

Take the job from the lathe and, using soft jaws in the vice, complete the tapping operation. Take care to keep the tap square, despite the guide, and use your best tap to finish with; you want a really good thread here. Use cutting oil and clear the tap frequently.

Make the screw next; this is in two parts, the thread proper, and the head. For the thread, simply chuck a selected piece of $\frac{1}{4}$ in. stock – a piece that is 0.250 in., preferably ground. With about $\frac{3}{8}$ in. projecting make the little boss. Get a good finish on the end, and aim at the shape shown. Draw the workpiece out of the chuck so that about $1\frac{1}{4}$ in. of $\frac{1}{4}$ in. diameter projects. Now, you want a really good thread, a sweet fit in the boss previously made. You may, perhaps, get a true thread with a tailstock dieholder alone, but I roughed mine out by screwcutting first. A 40 tpi thread is 0.016 in. deep, and I took out about 0.012 in. with a sharp screwcutting tool, simply to act as a guide for the die. This didn't take long, running at the slowest direct speed (225 rpm) and the long overhang from the chuck did not prove the problem I feared it might, taking out a thou at a time.

Next, fit up the tailstock dieholder, and adjust the die to the largest possible. With care and cutting oil cut the thread full length. Try it in the screwed boss. If it runs on, but stiffly, take another cut at the same setting, but if not, *very slightly* reduce the die diameter until you get a really good fit, with neither wobble nor backlash. This is very important, and if you get a sloppy fit, make another.

Now for the head. Chuck a piece of $\frac{3}{4}$ in. stock with about 1 in. projecting – enough to be able to part off to size. Clean up the end, and take a fine skim on the diameter. With a pointed tool make two grooves about $\frac{1}{32}$ in. wide, one $\frac{3}{8}$ in. from the face, and the other a further $\frac{5}{32}$ in. Set up a knurling tool and using oil and the back-gear knurl on the chuck side of the first groove. Don't run on to the parallel part between groove and face, but if you do, just turn it off. The diameter is not critical but deepen the groove again. Centre the end, and drill No.2, $\frac{3}{8}$ in. deep. As in the case of the boss, start the thread with a taper tap in the drill chuck, but in this case you can finish the screwing operation, using a tap-wrench, while the workpiece is still in the chuck. Let the

bottoming tap go in $\frac{1}{4}$ in. – 10 threads, no more.

The next step is to divide the parallel part into 25 grooves; 24 would do, but 25 is quite easy. (You can do 24 just using the chuck jaws, that's all.) There have been a number of articles showing how to divide using the changewheels and a detent, described both in *Model Engineer* and in many of the books published by Nexus Special Interests. Alternatively you can use a dividing head, if you have one, but for five or 25 divisions the leadscrew handwheel will do it for you. Set up as for screwcutting 8 tpi; then the mandrel and the leadscrew will rotate together at 1/1 ratio, and the divisions on the handwheel indicate the workpiece rotation. The only care to be taken is to avoid backlash in the system.

Grind a sharp-point tool of about 60° or 70° angle, and hone the end until you have a radius of about 20–40 thou wide. Set it sideways in the toolpost, first having honed it up to a really keen edge, so that it is exactly at centre-height, cutting face pointing towards the chuck. Angle it so that you have a top rake of about 5 degrees. Now set the changewheels or gearbox for 8 tpi, and get them in good mesh. Make sure that the direct drive pawl, if using an ML7, which engages with the bull-wheel, is also well in mesh. Slip the tailstock to the end of the bed, tie a piece of string to this, then twice round the leadscrew handwheel, and hang a heavy nut on the end or your riveting hammer. This will make a friction brake to keep the backlash all one way. Try a few practice runs without cutting. Loosen the headstock belts first, though you may find it easier to engage the back gear and use the belts to pull the mandrel round with. Pull the chuck over towards you until the index on the handwheel is at zero, then, always rotating the same way, turn until the index is at five, then ten, then fifteen, and so on, 25 times, and you should be back where you started. If your lathe has a 10 tpi leadscrew, then set up for 10 tpi, and go forward *four* divisions at a time.

Now for the cutting. Advance the tool until it touches and note the index reading. Withdraw the saddle to the right so that the tool clears the work. Set the chuck round until the leadscrew handwheel is at zero. Put on a small cut – say 2 or 3 thou – and wind the saddle chuckwards until the tool reaches the groove. Retract the cross-slide and saddle, put on another small cut and repeat. As you go deeper you will probably have to reduce the depth of cut. About 12 thou should be deep enough, and you will get a better finish if you repeat the final cut two or three times. The groove should then be about $\frac{1}{32}$ in. wide. Don't bother about the burrs at present.

Retract the tool fully, and then rotate the chuck to bring the index to 'Five'. If you go too far, go *well* back with the chuck and try again; it is imperative that you always approach the setting from the same direction, to avoid problems with backlash in the system. Repeat the shaping process and go on, right round the drum. It will pay you then to set on a *very* small cut – just a shaving – and go right round again, but this is, perhaps, aiming at perfection! It is a slow process, but it is surprising how fast the work goes once the initial try it and see period is over.

Once this is finished, remove the cord and the sideways tool and set up to machine the little bevel. Polish off all burrs with fine emery. Part off to size,

reverse, and grip very lightly in the chuck to finish the head. Remove from the lathe and degrease – I use carbon tetrachloride, but *Stergene* in hot water will do. Anoint both screw and head threads with *Loctite Studlock* and screw the parts together. Leave to cure. (If you have no *Loctite*, drill $\frac{1}{16}$ in. and fit a taper pin, filing the ends flush, not forgetting to ensure that any groove affected is still a groove!)

The slide-bar needs to be very rigid, as it is held rather a long way back from the operative end. The drawing shows a piece of $\frac{3}{4} \times \frac{3}{16}$ in. material, with a strong-back or elongated washer to give some stiffness. This served me well for some years on the ML7, but when transferring to the Super-7 (on which more later) I made a new one of $\frac{3}{4} \times \frac{3}{8}$ in. material, described on page 72. The slot can be milled out with a slot drill, but by the time you have set up for this you could just as quickly drill a row of holes and file it out; hand work is often quicker! The piece of angle at the end is either riveted or brazed; either will do (or both) but if you are using the $\frac{3}{8}$ in. thick bar, then a short piece of the same material brazed on to the end will be sufficient (see Fig. 3.16). You can braze up this part and the boss to the bracket now; I found my little *Soudogaz* butane blowlamp quite adequate for both jobs, but don't forget that you are working in steel; this means plenty of flux and heat as quickly as you can, or the flux will fail to cope with the oxide formed.

After pickling and washing, apply the 'finish'. The device will work as it is, but it *is* a quality product and should be treated accordingly. A good drawfile finish if you like (but no fancy curlicues or 'engine-turning' – quite inappropriate on workshop equipment to be *used!*). However, an attractive and functional

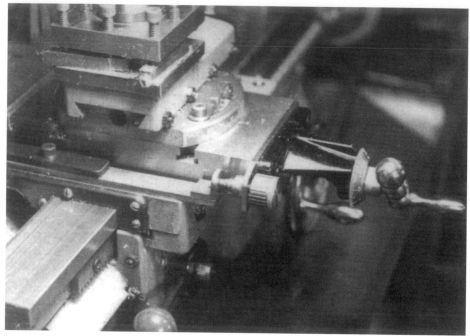

Fig. 3.14 Another view on the ML7.

finish is to heat-blue the job. This looks well and has the advantage of providing slight rust-resistance. Heat the whole gently and evenly, keeping the inner core of the flame well away from the metal. It will go through the 'temper colours' until it reaches blue, at which point dump it into engine oil or clean quenching oil, if you have any – though cooking oil will serve.

Offer all up to the lathe to see how it fits first, of course. You may find that the bar doesn't sit quite flat on the saddle; if so (and this applies especially to the thin type shown on the drawing) bow it slightly, so that when the fixing screw is tightened it flattens the bow. (The strong-back, which is a simple job to make, is also bowed with advan-

tage.) In fitting the bracket, part (4), you may find (on the ML7) that the cross-slide gib projects and fouls it. The handwheel bracket has a cavity to accept this. There is no alternative to removing the gib and shortening it as necessary, but this will give you the chance of taking off the topslide, cleaning out the underside, re-greasing the screw, and readjusting the slide when you put it back – all jobs which I do once a year, and the only way to ensure that all stays in order.

Now to make the detent, part (6). Here you will almost certainly have to experiment a little, as it is most unlikely that any spring steel you have will be the same as mine (1.6 mm wide × 0.25 mm thick). Find a piece as near as you can to this size; you may have to reduce the width. If so, then a saw-file (the type called a slim taper) will usually tackle the job, for springs are well tempered. Don't cut any off the coil but grip the end in a little hand vice so that about $\frac{3}{16}$ in. projects. Soften the end with a spirit lamp and then put a kink in it with a pair of pliers. If it breaks off try again, but this time get the spring a bit hotter. The less heat you apply the better, and it is more effective if you have to make three to get the last one right than to overheat the first one and have it fail in service. Don't cut it off yet.

Find, or make, a little 4 BA screw as shown on the drawing. It should be one with a fairly tall head and should be steel. Put a slot in the end as narrow as you can manage – you may have to grind the 'set' off a 6 in. mini-hacksaw blade, but I have an *Eclipse* jeweller's hacksaw which is worth its weight in gold for this sort of job. Remove any varnish and, if necessary, the tempering colour from the spring and tin it with a small, hot, soldering iron. Enter it to the slot with the screw held in place on the plate and adjust the length until the kink engages nicely with the slots in the drum; then remove the screw and solder the spring in place with tinman's solder, not the soft resin-cored stuff. Note – if the spring is a bit slack in the slot pack it with brass shim, don't try to make it good with thick solder films. Wash well, and then cut off the surplus spring.

Offer up to the drum again and mark the exact spot for the tapped hole. Drill and tap 4 BA (after removing the drum, of course). Fit a locknut on to the detent thread and screw the whole in place with the spring about $\frac{5}{16}$ in. from the face of the bracket, part (4). Put back the stop-screw (5) and then adjust the detent until the drum can be rotated without too much resistance but is, at the same time, held firmly by the spring. Tighten the locknut. The job is now ready for work.

Mark II as fitted to the Super-7 lathe
Before telling you how to use the stop, it might be worth while saying a little about the device as fitted to my Super-7. When the ML7 was sold, the stop then fitted went with it. I did not want to lose any of the travel of the cross-slide, so instead of fitting a plate to carry the stop-screw I bolted a small block to the side of the cross-slide, as seen in Fig. 3.15. This is $\frac{3}{4}$ in. square × about $\frac{5}{8}$ in. long and attached with two 2 BA Allen screws (see Fig. 3.16). The only other change is in the length of the stop-screw itself, which is made so that the screwed portion projects from the drum a shade over $\frac{1}{8}$ in. This provides a working travel of $\frac{1}{4}$ in. which is enough for the largest screw likely to be met with.

There is no adjustment for thread clearance on this one, so that more care must be taken with the fit of the $\frac{1}{4}$ × 40 screw. Note that I mean *fit*; this does not mean using a smaller tapping drill, but rather to take more care to ensure that the threads are well formed. The tapped hole in the block is made progressively i.e. the taper tap is taken part through, followed by the seconds, then the 'plug'. The taper is then put through further and the process repeated until the hole is tapped right through. A tapping oil is used, such as *Rocol Ultracut*, and the chips cleared frequently. When cutting the male thread I started by screwcutting to about $\frac{3}{4}$ full depth and then followed with the die, adjusting this as described

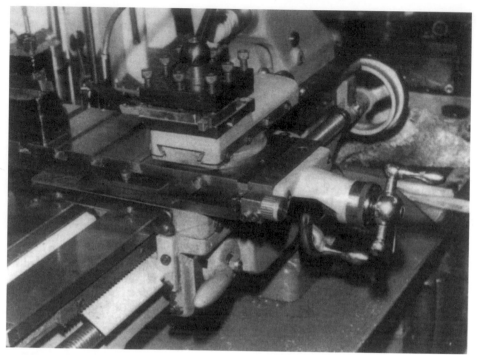

Fig. 3.15 Mark II design fitted to a Super-7 lathe. The square-headed screw should be replaced by a socket grubscrew.

already until the thread was a sweet fit in the hole.

Once the block was made it was clamped to the cross-slide and the holes spotted through, after which the holes were drilled and tapped $\frac{3}{8}$in. deep. I did not want to break through on to the dovetail shear of the cross-slide. The slide bar, Fig. 3.16, is much stiffer, giving a more positive 'feel' in action.

Method of use

Whether the bracket or the side block is employed, either can be left in place – they don't get in the way, but I find that the slide-bar assembly can be a nuisance at times. It is up to you whether you leave it there or fit it only when needed, for it takes but a second or so. The square-headed screws seen in Fig. 3.15 (which retain the topslide) are best replaced by $\frac{3}{8}$in. BSF socket grubscrews. After setting up the device, the index drum is screwed right forward and then back about six clicks. The tool is advanced to within ten thou of the workpiece and the slide-bar then brought forward until it touches the stop-screw. Tighten the Allen screw – it should be tight, but don't overdo it and strip the thread! Start the lathe and with the cross-slide against the stop adjust the drum until the tool just touches the work. You are now at 'zero' cut.

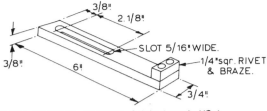

Fig. 3.16 Details of the Mark II design of depth stop.

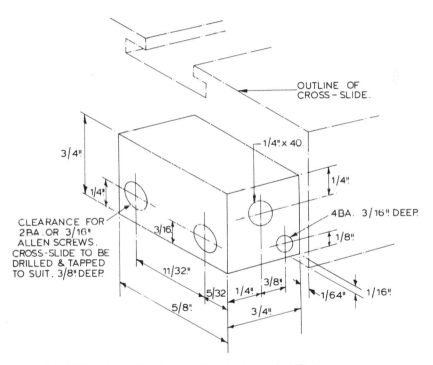

3/8"

2. 1/8"

SLOT 5/16" WIDE.

3/8"

6"

1/4"sqr. RIVET & BRAZE.

3/4"

SLIDE BAR : M.S. Replaces Original part N° 1.

OUTLINE OF CROSS – SLIDE.

1/4" x 40.

3/4"

1/4"

1/4"

4 BA. 3/16" DEEP.

CLEARANCE FOR 2BA. OR 3/16" ALLEN SCREWS. CROSS-SLIDE TO BE DRILLED & TAPPED TO SUIT. 3/8" DEEP.

3/16"

1/8"

11/32"

5/32

1/4"

3/8"

1/64"

1/16"

5/8"

3/4"

BLOCK : M.S. Replaces Original part N° 4.

Adjust the knob of the stop-screw to apply the cut you need; one click is one thou. Advance the cross-slide to the stop and start screwcutting in the ordinary way. At the end of the cut withdraw the cross-slide, disengage the clasp nut and retract the saddle. Adjust the drum as many clicks as you need thous of cut and advance the cross-slide to the stop, engage the clasp nut and make another cut, and so on. You don't have to look at the cross-slide index at all unless you have to use it to determine a definite depth of thread,

73

and even then you only have to keep an eye on it every now and then. The drum looks after the depth of cut at each pass. At the beginning of each pass the cross-slide is brought into contact with the stop; at the end it is retracted as much as you wish to clear the thread already cut. You will be agreeably surprised at how easy it is to use and how quickly you can cut a thread once you have got used to it!

I do use the device for normal turning when there is a lot of repetition work and if you think a little you will realise that it would have simplified the cutting of the grooves in the drum to even depth no end. It can also be used for internal threads. The obvious way is to fit the measuring drum at the other end of the cross-slide and reverse the slide-bar, but this is awkward to use and you would need two drum assemblies, as otherwise the time taken to transfer would make the exercise hardly worthwhile.

But there is another way, which has decided advantages and that is to turn the screwcutting tool upside down and work at the *back* of the hole. The operation is then exactly the same as for a male thread. There is the further advantage that all the movements are exactly the same as for external screw-cutting – you don't have to think back-to-front, and you can see what is going on much better. The tool can be held in the normal four-tool turret if the hole is not much more than an inch in diameter, or you can mount it in the rear toolpost. Parting-off tools have been used in this fashion for a long time – it is odd that no-one has thought of screwcutting (or even boring) with an inverted tool as well!

Slotting operations
The device can be of real help when slotting (e.g. keyways) in the lathe. The normal procedure of reciprocating the tool with the saddle results in rubbing during the back (idle) stroke, causing wear on the clearance edge and virtually blunting the tool. However, if the slotting tool is carried in the *rear* tool-post the depth stop can be used in the manner already described for internal screwcutting. This avoids all rubbing, and it will be found that the cutting action is much sweeter. There is one further point worth mentioning – it will pay to take *two* cuts at each depth setting, as there is usually some whip in the necessarily slender tool.

Using 8 mm watchmaker's collets in a No. 2 Morse socket
(*Based on an article which first appeared in Timecraft*).
The collet is the Rolls-Royce of work-holding devices but unfortunately those for lathes of any size are prohibitively expensive for most people. However, the change from mechanical watches and clocks to quartz crystal timepieces has meant that 8 mm collets are appearing for disposal fairly frequently – even new ones are relatively cheap. True, they have a limited capacity – about 6 mm is the limit – but within that capacity they offer a very accurate method of workholding. I have two adaptors which enable 8 mm collets to be held directly in the No. 2 Morse taper socket of the Myford. One is by Wolff-Jahn, which has a No.1 MT and has to be used with a sleeve; the other is home-made. This one is easy to construct and though, when first made, it had the disadvantage that it had to be removed from the mandrel to release the collet, this can be overcome. See Fig. 3.17. It was, in fact, made not so much to hold workpieces as to carry

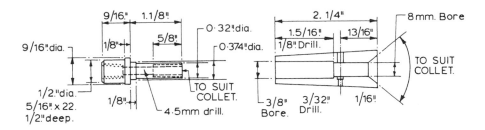

Fig. 3.17 *Adaptor to accommodate 8 mm collets within a No. 2 Morse taper socket.*

small 8 mm arbor-mounted cutters when milling in the lathe. It was no problem to alter the design to enable it to be used with a normal draw bar release from the tail of the mandrel, and what follows allows for this.

The basis is a No 2 MT Jacobs chuck arbor; almost any size will do, but I advise those which have a moderately large chuck end, say Jacobs taper No 6 or thereabout. These are of carbon steel, but not hardened, and with care and sharp tools a good finish can be obtained. I recommend the use of a soluble (water diluted) cutting oil, but a straight cutting oil like *Macron B* will serve. A good finish is important, of course, especially in the bores.

The main problem is to hold the thing, for it is very difficult if not impossible to obtain the enlarged bore at the back unless the work can be tackled from both ends. The first operation, therefore, is to set the arbor between centres and machine down both ends as shown in Fig. 3.18. The diameter at the large end (which will later be held in the fixed steady) I recommend should be not less than 0.475 in. Now set the workpiece in the lathe mandrel socket and machine off all but $\frac{1}{4}$ in. of the cylindrical end, and form a new, deep, centre. Remove from the lathe and saw off most of the tang – leave just enough

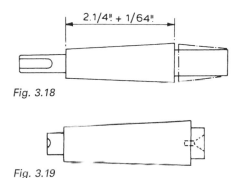

Fig. 3.18

Fig. 3.19

of the cylindrical part to carry in the steady; say $\frac{5}{16}$ in. (Fig. 3.19).

Set up the large end in the four-jaw chuck and centre very carefully, using a dial indicator, with the other end equally truly set in the fixed steady. This can best be done by setting a dial indicator on the free end before fitting the steady, and then using the cigarette paper technique to adjust the jaws of the latter. Face the end, centre, and drill $\frac{9}{32}$ for 1.375 deep. Bore the hole 0.375 in. dia, full depth. This may be hard on such a tiny boring tool, and my own method was first to use the boring bar to true the hole, then follow with progressively larger drills up to $\frac{1}{64}$ in. under size and finally bore again.

Set the arbor back in the mandrel socket, drill (say $\frac{9}{32}$ in.), bore truly, and

75

finally ream $\frac{5}{16}$ in. This hole must be made in two stages, as it must be mounted on a mandrel between centres to form one face. Do this next. Cut off all surplus at the small end with a saw, mount on a mandrel as explained and face the end. A small countersink can be made by hand afterwards. Back to the lathe mandrel socket, face the end, bore 8 mm making this a sweet fit to your particular collets. They do vary from make to make by a fraction of a mm, and an over-tight collet can be a nuisance.

You must carefully check the angle of your collets before forming the conical socket. What I did was to machine the cone almost completely to 40°, and

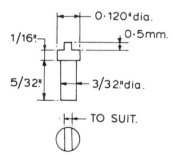

Fig. 3.20 Peg to prevent rotation of the collet in the socket.

then checked with a solid collet arbor and marking blue, adjusting the set-over of the topslide until I got at least 75% transfer of colour. Finish machine the socket until the head of the collet just sinks flush when gripping a rod of the correct diameter. Carefully deburr all sharp edges. Incidentally, if you are not happy about the finish on the cone, don't polish with emery. Rehone the tool, getting it really sharp (I use an Arkansas stone on mine) and take a fine shaving cut to rectify matters.

Whether or not you fit a locating dowel is a matter of choice. I have another machine which has none, and only the occasional collet 'acts awkward' and turns in its seat. However, I show (Fig. 3.17) a simple way to fit one. The $\frac{1}{8}$ in. hole is drilled from one side, right across the bore, and penetrates about $\frac{1}{16}$ in. *across* the bore. This is followed by a $\frac{3}{32}$ in. drill. The little peg can now be made – I use silver steel but don't harden it – and set in the $\frac{3}{32}$ in. hole quite easily. You will find you have to adjust the dimensions a bit to suit your collets. See Fig. 3.20.

The drawhead (Fig. 3.21) is made from mild steel, not carbon steel, as I prefer this to give way rather than the thread on the expensive collets. Drawn gunmetal will do just as well, and be

Fig. 3.21 The collet-tightening rod. Note this is not intended as a draw bar (see text).

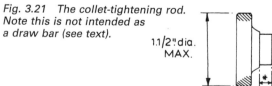

* TO SUIT LATHE . SEE TEXT.

even kinder to the threads. It is turned down from $\frac{5}{8}$ in. dia stock. Grip in the three-jaw, about $1\frac{1}{4}$ in. projecting, and chew off surplus metal down to $\frac{7}{16}$ in. dia just under $1\frac{1}{8}$ in. long. Face the end, centre, and drill tapping size for your particular type of collet, $\frac{5}{8}$ in. deep. This may present a problem! Table 1 (page 79) gives the threads of several of the more common makes, but is by no means exhaustive. I may say that I have found that 40 threads/inch draw bars fit all my own collets, even though some are 0.625 mm pitch; the difference is only 4 tenths of a thou per pitch. Where I do have the proper tap, I still screwcut 40 threads/in. and follow with the 0.625 mm tap; not, perhaps, the best tool-room practice, but it works! Those with metric lathes can, of course, screwcut properly – their problem comes if the pitch is 40 threads/in.

Expanding a little on this matter, the chief difference between these threads is the thread angle; those at 40 threads/in. will be 55° Whit form, the 0.625 mm threads will be the metric 60° shape. The difference in o/d between the largest and the smallest collets is only 0.007 in. and is unlikely to cause problems.

The tapping size for either pitch should be $\frac{1}{2}$ mm smaller than the thread o/d; this will give about 64% of full thread engagement – quite sufficient for this application. If you have the correct tap, there will be no difficulty, otherwise you must screwcut the hole until the collets fit sweetly.

Try more than one, especially if they are old, as I find they do wear in time. Lightly countersink the end of the hole, then follow with a 4.5 mm drill – this is the largest size you can pass right through an 8 mm collet – about $1\frac{1}{4}$ in. deep. Next, finish turn the o/d and the face which will bear on the end of the adaptor. Again, aim for a good finish, as friction here will reduce the grip of the collet. The $\frac{3}{8}$ in. dia shoulder should be a smooth fit in the hole in the adaptor.

Reverse in the chuck (after sawing off any surplus length) and machine the head. Note that this head must fit within the taper socket of the lathe mandrel, so that the sizes shown should not be exceeded. If you wish to operate the collet from the tail end of the mandrel, then you must now drill and tap for the operating bar, so drill and tap $\frac{5}{16}$ in. × 22 (BSF) or × 26. Remove all burrs. On the other hand, if you are content to take the whole thing out to change workpieces, all you need is a cross-hole for a tommy bar.

The operating rod is shown in Fig. 3.21. The lengths marked * must be taken from your particular machine. That shown as d* should be a good, but not tight, fit in the hole in the end of the mandrel so that the whole does not wag about. Note, however, that this is not a draw bar, and the face of the knob should not butt against the end of the mandrel. I have shown the knob as $1\frac{1}{2}$ dia, and this should not be exceeded. If anything it is a bit large as you don't want to over-tighten collets in use.

Check that the length is correct and will draw up the collet without fouling the end of the mandrel, and then screw the operating rod firmly into the draw-head; use either *Loctite* or a $\frac{3}{32}$ in. pin to fix it. You will appreciate that when assembling the parts on the machine the adaptor must first be set in the lathe mandrel; then the drawhead and oper-ating rod must be threaded through from the tail end of the lathe mandrel. This leads to an important point. It is vital that you thoroughly clean out the mandrel bore before setting up, otherwise there is a risk of dirt getting into the threads or, worse (if you change collets) of dirt getting onto the collet seat.

The device has proved to be very accurate in service and has enabled me to transfer work from the Boley to the Myford and back with no loss of concentricity. The only snag I have found is that the knob does foul the changewheel guard – not important if you have a screwcutting gearbox, but a nuisance if you rely on changewheels. The solution is to work with the guard off when using collets.

Wolff-Jahn adaptor
Figure 3.22 shows the design. It is actually for a No. 1 MT, but I use it with a converter socket. It has the advantage (a) that the collet grip is external and (b) it can be used with a draw bar (though the hole had to be screwcut first – it had no thread when I first had it). The defect is the fair amount of overhang from the mandrel nose – almost $2\frac{1}{4}$ in. when used with the reducer socket. As I have not made it myself I will give no description of how to do so, but if I *were* going to make one I should start with a No. 2 MT arbor as before, and use a fabricated construction. The end of the arbor could be threaded say $\frac{1}{2}$ in. × 26 threads/in. into a 'tee' of $\frac{3}{4}$ in. square material, and the nose likewise, the threads fairly slack and the whole brazed up solid with silver solder. Note, however, that it would be prudent to drill the Jacobs arbor and thread for the draw bar first, as there is just a faint risk of it going hard on you when brazing. I know it *shouldn't* happen, but it does! The o/d and the bore, together with the conical seat, can all be machined with the

rough 'fabrication' set in the lathe mandrel socket; the cavity for the tightening screw has, on mine, been slotted out with a slot drill, and the concave sides have probably been cut with an endmill before polishing.

I don't use this one in the lathe, following an unfortunate impact between a finger and the rotating tee, but I do use it in the milling machine to hold small cutters mounted on 8 mm arbors. One does not (if fit to use a milling machine at all) bring fingers close to the revolving quill! It is a handy little gadget, and, as one would expect from this manufacturer, accurate and well finished.

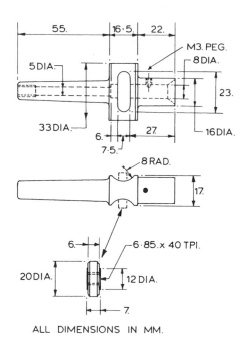

ALL DIMENSIONS IN MM.

Fig. 3.22 The Wolff-Jahn collet adaptor, in this case for No. 1 MT. Dimensions in mm.

TABLE 1 Threads on various 8 mm collets

Make	Dia.	Pitch (mm) or threads/in.
GeoAdams	0.268*	0.625 mm
Boley	0.268	0.625 mm
Boley-Leinen	0.268	40 t/i
Coronet	0.268	0.625 mm
Derbyshire	0.268	40 t/i
IME	0.275	0.625 mm
Lorch	0.275	0.625 mm
Pultra	0.268	0.625 mm
Wolff-Jahn	0.270	40 t/i

Stated as 0.270 in. (6.86 mm) in some sources.

Collet keyways

Make	Width (mm)	Depth (mm)
Boley & Leinen	2.0	0.5
GeoAdams	1.75	0.7
Lorch	1.75	0.65
Wolff-Jahn	1.875	0.7

A headstock length stop

There are many methods of setting-up to part off successive pieces of stock to dead length, the easiest, perhaps, being to use a pad fitted to the tailstock barrel against which the bar can be set before parting off. However, there are cases where this cannot be done, and it is of no help when stock already in short lengths must be cut to exact size. Internal length stops are common accessories for collet-type headstocks. This one is of similar design, and was made to ease the machining of a number of columns for a marine engine.

The length bar is held inside the hollow headstock mandrel by means of an internal and external expanding taper, which grips both the mandrel and the internal length bar. The latter is supported in the centre of the hole by means of a collar. Normally the bar is used alone, but if (as in the case of these columns) there is a screwed peg projecting beyond the shoulder to which the 'exact' dimension applies, then a socket-pad is attached to the end of the bar.

Fig. 3.23 shows the arrangement. The inner cone 'A' is made from silver (carbon) steel hardened and tempered, but mild steel will serve for the remainder. The dimensions shown are those for the Myford Super-7, but should also fit the ML-7 series. Note that though the lathe specification shows the spindle to be bored to pass $\frac{19}{32}$ in. the greater part of this bore is 0.625 in. dia – this should be checked before making the fittings.

The outer cone 'B' should be made first. I used hex steel stock to fit a $\frac{7}{16}$ in. Whit. spanner, and the same size served for the nut 'C'. I don't advise the use of knurled hand knobs here as though they look nice they don't give sufficient grip.

Chuck the stock with a couple of inches projecting and turn the o/d to the dimensions shown, at the same time bevelling the edge of the hexagon. Centre and drill $\frac{7}{16}$ in. Now set over the topslide to 3° (the taper is not critical) and bore the cone until the diameter at the large end is $\frac{9}{16}$ in. Take care to get a very good finish from the final cut. Remove the internal and external sharp edges. Leave the topslide set to this angle until the mating cone has been made. Draw out from the chuck and part off. While the hex. stock is still in the chuck you can make the nut, part 'C'. Face and machine the bevel, then drill $\frac{13}{32}$ in. and tap $\frac{7}{16}$ in. × 32 (or use $\frac{7}{16}$ in. × 26 if you have none finer, adjusting the tapping drill to suit; don't use 40 tpi, as

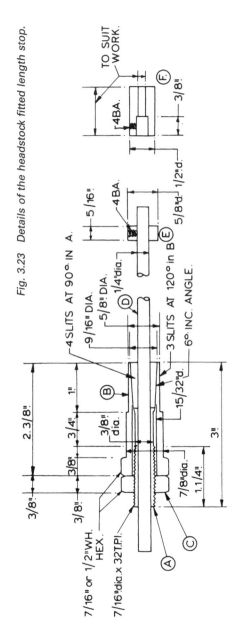

Fig. 3.23 Details of the headstock fitted length stop.

this is too fine for such an application). Part off, face the other side and make a bevel there, too. Return to part 'B'; grip by the cylindrical part and either drill or bore to $\frac{15}{32}$ in. dia for $1\frac{3}{8}$ deep i.e. as far as, but not running into the taper. Face the end, put on the bevel and remove all burrs.

The inner cone, part 'A', is a bit of a pig to hold, but the following procedure seemed to work well enough. I had a piece of $\frac{9}{16}$ in. dia silver steel available, otherwise I should have turned down from $\frac{5}{8}$ in.; you will need a piece at least 4 in. long ($\frac{9}{16}$ in. stock will, of course, pass through the headstock bore). Centre the end and drill reaming size for $\frac{1}{4}$ in. and follow with the reamer, running slowly and with plenty of cutting oil and frequent withdrawals for chip clearance. Bring up a large Slocombe drill and put a slight cone in the mouth of the hole. Draw out about $3\frac{1}{4}$ in. from the chuck and bring up your tailstock to support the end.

With your parting tool make grooves $\frac{1}{16}$ in. deep to coincide with the end of the taper, the forward end of the thread, and the 3 in. overall length. The depth isn't critical at all, but better a few thou deeper than too shallow. Now turn the taper, again taking great care with the final finish; polish with emery if necessary. The taper should just run out at $\frac{9}{16}$ in. dia at the large end – let the other end look after itself, but not more than 1 in. long. The next step is to reduce the diameter between the two grooves to $\frac{3}{8}$ in. – a good tool finish is all that is needed here – and then machine the remainder to $\frac{7}{16}$ in. dia. Set up the changewheels for 32 tpi, and screwcut the thread. Now, you can't test this against a nut, and you can't hold the work the other way round to use the tailstock dieholder either. So, carefully stone the tool to get it really sharp and the correct angle (55°) and very slightly round the point. After setting to the work with your gauge, bring the tool up until it *just* brushes the o/d and set your cross-slide index to zero. (You should, of course, have reset the topslide, too.) Screwcut carefully, adjusting the top-slide one thou to the left for each 3 or 4 thou you feed in – this eases the cutting action. Carbon steel is not easy to cut with no top rake on the tool (and all my screwcutting tools have zero rake) so don't use too much infeed at a time and use lashings of cutting oil. Full thread depth is 0.020 in., so stop when you have fed in 0.018 in. – the rest can be removed with a die in the vice later.

Part off, reverse in the chuck and grip *lightly* by the thread, with the cone within the chuck jaws. Put a $\frac{17}{64}$ in. drill down 2 in. deep – this leaves only the bore under the cone reamed to exact size. Round the end of the thread and take off the sharp edge from the hole. Try on the nut, and if need be take off a little bit at a time with a die; don't do too much at once, as the workpiece isn't all that strong at the centre recess. The nut should run on easily but not be sloppy.

Part 'E' should need no instructions! The o/d should be an easy but not 'rattly' fit in the bore of the headstock. The Allen grubscrew can be any convenient size – mine is 4 BA – but the 'point' should be ground off so as not to burr up the bar 'D'. (The latter is no more than a piece of $\frac{1}{4}$ in. BDMS, and I have three, of different lengths from 15 in. to 4 in. long.) The recessed button, part 'F', is made to suit the job in hand; the dimensions shown are for a 4 BA stem projecting from the shoulder of the columns.

The collet cones must now be slit.

Note that I have shown three slits in the outer cone, and four in the other one. You can use four in each if you like. I set out the holes at the end by eye, drilled $\frac{3}{32}$ in. at the inner end, and then slit down with a small hacksaw. No doubt the purists would object and call for the use of slitting saws, but 6 in. hacksaw blades are cheaper and quicker! Which-ever way you use, file off the burrs inside and out with a fine three-cornered file and then go over the whole and again remove all sharp edges.

The inner cone will work well enough without hardening, but will last longer and work better if heat treated. Wrap a bit of iron wire round the thread and coat this well with a mixture of pow-dered chalk and water. Set to dry, assisting the process with gentle heat if you like. This will reduce the amount of scale on the threads. Heat to cherry red and hold this temperature for a couple of minutes before plunging vertically, cone downward, in oil. I have the proper quenching oil but thin motor oil will serve for this job. You *can* quench in water, but it is just a bit fierce for this application; if you have no oil, use water heated until it is just too hot to bear your hand in (that is about 150°F). When quenching, move the work about *gently* in the oil or water; this is more effective than rapid swir-ling about.

Dry off and then polish the conical end. Heat gently by the threaded part until the cone turns purple-blue and quench in water or oil. The cone may be left in this condition – indeed, I blue many of my tools of this sort, both to reduce risk of rust and for appearance sake.

Clean up all parts, and assemble the collar 'E' on the stem 'D'. Assemble 'A', 'B', and 'C' hand tight and slip the stem

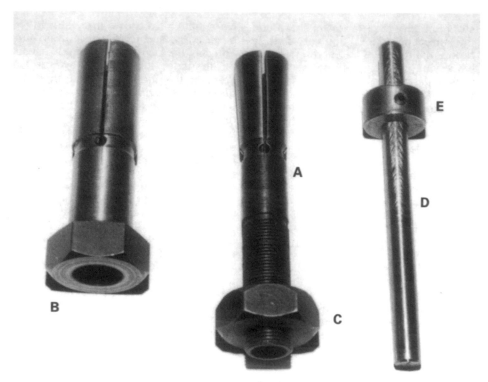

Fig. 3.24   Parts of the length stop.

through. Enter into the headstock mandrel and tighten up the nuts – you may need two spanners at first. After a while you should find the collet set gripping the bore of the headstock mandrel but the rod 'D' can move. Adjust the rod and then tighten still further, until a light tap on the end causes no movement. This is tight enough. To remove, slack off the nut, and give it a tap inwards. This will free the two cones and the whole can be withdrawn. Fig. 3.24 shows the parts.

In use, of course, the final, precise, process of setting to length is done with the lathe index dials either on the topslide or the leadscrew handwheel.

In the case of the engine column I mentioned (Fig. 3.25) the shortest of these was measured between the shoulders using vernier calipers and fitted with the appropriate attachment part 'F' with about $\frac{1}{8}$ in. projecting from the chuck (to the shoulder, that is). The measured amount was then taken off using the topslide index, and the tool then locked. After a check measurement on this column, the other seven were then machined in turn. A check showed that all were within half a thou up or down compared with each other i.e. the maximum difference between lengths

Fig. 3.25 An example of the work for which the stop is used in bringing the length between shoulders to within acceptable limits.

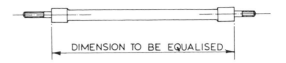

DIMENSION TO BE EQUALISED

was just under one thou and all were within this tolerance of the designed length. This I regard as satisfactory, especially as measuring to such small shoulders with a vernier is subject to slight errors.

This method will not, of course, serve for short workpieces, as the stem cannot enter the chuck jaws in many cases. For such work I use a 'top hat' fitment shown in Fig. 3.26. I have a few of these, made as the need arose, and Fig. 3.26 is one which served when I had a multitude of little collars, $\frac{1}{4}$ in. dia and in lengths ranging from $\frac{3}{8}$ in. to $\frac{7}{8}$ in., about a dozen of each to make. Note that this device is *not* intended to hold the work while turning on the cylinder – there is always a little run-out. Its purpose is to hold for machining to length only, it is not a substitute collet. Most of those I have made have no adjusting screw – Fig. 3.26 is an elaboration for the special case just mentioned.

Finally, for those with collet-type lathes a device similar to Fig. 3.23 can be made up to fit inside even an 8 mm

collet shank. The outer diameter of the adjusting screw must be small enough to pass inside the draw-tube, and (on mine, anyway) both the 'nut' and the outer cone are fine knurled. The core diameter of the threads of such (8 mm) collets is about 0.235 in., so that the o/d over the knurled nut should be a few thou less than this. The design is very much the same, the taper being of the same order on the cones, and the rod or spindle is about 2 mm dia. In such a device I would recommend hardening and tempering all parts.

A milling spindle drive or overhead

There is no more versatile machine tool than the centre lathe, even in its basic form. With the addition of a vertical slide it can become a small milling machine or even a jig-borer of tolerable accuracy, while the accessories used by the ornamental turner enable work of astounding complexity to be done. These, in their modern form i.e. those built since about 1825, depend a great deal on the use of drill spindles and cutting frames held in the slide-rest and driven from an overhead gear, the headstock mandrel being stationary or being used only as an indexing head. There have been many designs of such accessories over the years, although that due to the late Mr Potts is perhaps the best known. Indeed, the Potts spindle is almost a generic name, irrespective of make.

Such devices need a separate drive, and there are many types of these also. Some, like the Pittler, incredibly complex,

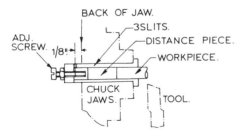

Fig. 3.26 A headstock length stop for a short workpiece.

83

while others are simple in the extreme, with consequent defects in service. Many incorporate a long drum mounted on standards above the lathe bed, the idea being that as the cutting frame is traversed so can the drive band follow it. Unfortunately a jockey arrangement is needed as well, to keep the band tight, and this – almost without exception – does not or will not traverse with the cutting frame. Others rely on complex systems of pulleys, again with the idea that the final pair can swivel and so follow the travel of the slide-rest. I have had many types on various ornamental lathes I have owned from time to time, and although all 'work', none was very satisfactory.

The original design of my lathe bench provided for a substantial overhead drive, with its own motor, all disposed on a shelf above the lathe. For various reasons this was not proceeded with, and just as well for as the years passed and I gained more experience with those on my ornamental lathes I realised that something much simpler could be devised. The requirements were:

(a) minimum number of jockey pulleys
(b) quick to set up and take down
(c) not to be too expensive
(d) not to lock up a motor which would spend most of its time stationary
(e) one which would face all ways as needed.

The nearest to this had been the 'ship's davit' type as fitted to Holtzapffel lathes, but the shelf above the machine would have got in the way. However, it did not seem impossible to use this as a basic idea. A further point I had found was that with foot-mounted motors of about $\frac{1}{4}$ hp there was no need to bolt these down – their own weight was sufficient to take the very light drive needed for quite hefty cutting frames. Little rubber feet prevent any 'walking about' and on my larger Holtzapffel the cutting-frame motors sit quite happily on a polished mahogany back shelf.

The final design comprised a vertical standard made from electric conduit secured in the Myford hand-rest bracket which can be clamped to the bed in any position, or to the cross-slide. To the upright is attached a horizontal arm which can be rotated through 360° and, of course, be set at any height. Two deep-groove jockey pulleys can be slid along the arm. Thus the assembly can be set anywhere on the lathe bed not occupied by the saddle (or tailstock) or on the saddle, and the pulleys can face in any direction; *and*, there are only two pulleys! Experiment showed that the motor could also be set almost anywhere – behind the lathe, at the end, or even on the lathe bed if necessary.

The standard (Fig. 3.27) was made from $\frac{3}{4}$ inch o/d conduit, the end being turned down for about 2 inches to fit the hand-rest base. The side arm, also seen in Fig. 3.27, is $\frac{1}{2}$ in. dia steel and brazed. (The drawing shows a plain spigot which would be adequate.) Buttons are turned to fit into the ends of the conduit, decorated to taste. The jockeys, Fig. 3.28, were made from aluminium alloy, although cast iron is preferable provided it is bushed with gunmetal; this will better stand the wear. However, you may not have any stock that size so why not use wood? My Holtzapffel No. 484 has the overhead pulleys of wood, the former box and the latter mahogany. They have lasted for nearly 200 years! But the best species for this application would be lignum vitae, and this can be obtained from a worn-out bowls wood provided it is not too

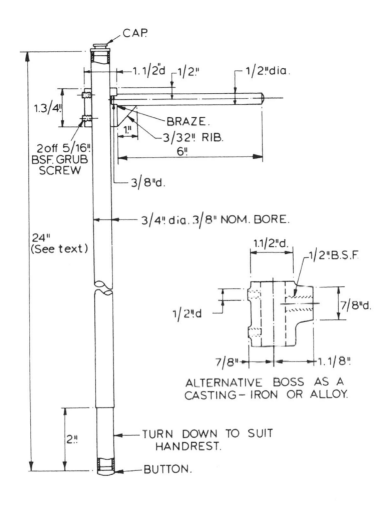

CAP.

1.1/2"d

1/2"

1/2"dia.

1.3/4"

2off 5/16"
BSF. GRUB
SCREW

BRAZE.

1"

3/32" RIB.

6"

3/8"d.

3/4" dia. 3/8" NOM. BORE.

24"
(See text)

1.1/2"d.

1/2"B.S.F.

1/2"d

7/8"d.

7/8"

1.1/8"

ALTERNATIVE BOSS AS A
CASTING – IRON OR ALLOY.

2"

TURN DOWN TO SUIT
HANDREST.

BUTTON.

COLUMN AND ARM.
B.M.S. Throughout.

Fig. 3.27

badly split. However, almost any wood will do, and the procedure is to drill the blank say $\frac{5}{8}$ in. dia and *Araldite* in a length of brass or gunmetal projecting about an inch. This is then your chucking-piece for the turning operation, and can be sawn off and faced afterwards. The only point to make in this simple turning operation is to make sure that the diameter is at least 8 times the belt diameter; the drawing is right for $\frac{1}{4}$ in. leather. The bore should be an easy

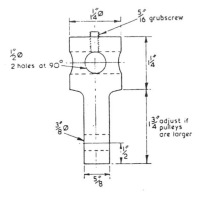

PULLEY BLOCK 1 off
al. alloy or B.M.S.
(or al / C.I. casting)

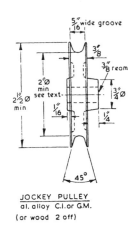

JOCKEY PULLEY
al. alloy C.I. or G.M.
(or wood 2 off)

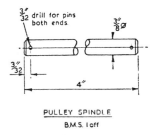

PULLEY SPINDLE
B.M.S. 1 off

Fig. 3.28 Jockey pulleys and hanger.

running fit to the spindle, which is bright drawn steel. Alternatively, of course, you can fit small ball bearings.

The pulley hanger can be made from $1\frac{1}{4}$ inch square or round material – steel or aluminium alloy – and calls for a fair bit of hacksaw work (or you can mill it out if you like) but is not at all critical as to dimension. You will see that it has two holes at right angles to fit the supporting arm, so that it can be set up either way round. The pulley spindle is secured in the block with *Araldite* or *Loctite* or you can use a setscrew. It should not be free to move.

Having read so far you will wonder how to machine the foot of the standard when your lathe is only 19 inches between centres. Remove the tailstock, set the tube in the three-jaw and adjust the fixed steady to it. Move the steady to the end of the bed and re-chuck the tube, holding about $\frac{1}{2}$ inch in the chuck jaws. Machine a 2 inch length to as close to the jaws as you can get, and then part off. Remove burrs, and there you are!

Now a word about belts. I use leather, joined with a proper bent-wire fastener as on the old-fashioned sewing machines. This is still obtainable from some sewing machine shops but if in difficulty ask a saddler to get you some. He will also be able to supply you with a little neatsfoot oil in which you should soak the leather for an hour or so, wiping off the surplus afterwards. This will make it supple. For belts $\frac{1}{8}$in. dia or below the screw hook-and-eye type of fastener is best, and these are not always easy to get. Both Buck and Hickman and Buck and Ryan used to stock them, as did the sewing machine people or try horological suppliers. The best joint, of course, is one which is glued and sewn and again, the saddler can do this for you once you have decided the right length.

I have given up using plastic belt, as it tends to whip a lot, though I do have some heavy ribbed material which is not too bad. All plastic belt must be pre-stretched by hanging a weight on the end overnight. The joint in this case is made by heating the ends with a hot knife and welding it. A third alternative is braided nylon cord, as used for starter cords on hovermowers and chainsaws. Make sure it *is* nylon, though. This can be welded too, by teasing out the ends and after twining the two ends together, heating in a match flame or small spirit lamp. The drawback with this stuff is that although it is very strong it has practically no 'give' and is difficult to adjust. Finally, those of you with Navy service will, of course, make a long splice in a piece of cotton cord, and have the best drive belt of all!

Now for the application. First, though, note that the normal Potts spindle has a pair of jockeys of its own. In some cases these can help, but the set-ups shown in my photos are deliberately made without them, to give some idea of the flexibility of the device. Fig. 3.29 shows the motor on the lathe bed standing on a piece of wood with the Potts spindle set up as if to drill a ring of holes on the face of a disc. The spindle was running at 1500 rpm and was quite satisfactory with both leather and braided nylon cord. Fig. 3.30 shows the identical set-up of the Potts, but with the motor behind the bed. In both these cases it was possible to traverse the saddle about $4\frac{1}{2}$ inches,

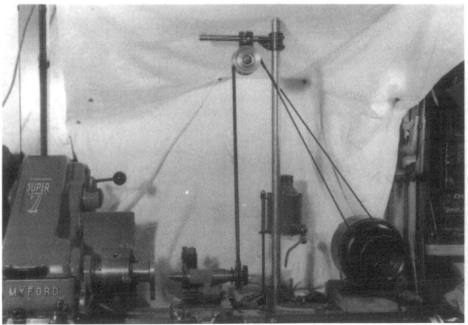

Fig. 3.29   A Potts spindle driven from a $\frac{1}{6}$ hp motor on the lathe bed.

Fig. 3.30 *Spindle set up as in Fig. 3.29 but with the motor behind the lathe.*

but had the standard been fitted to the saddle itself the second set-up allows up to 7 inches travel.

Fig. 3.31 shows the Potts carrying a milling cutter, attached to the vertical slide, the standard at the front of the cross-slide and the motor again behind the lathe. At above 8 inch travel of the saddle there was some slackening of the drive. This did not matter with leather, but braided nylon lost its grip. This was a situation where the more 'stretchy' plastic belt might have had an advantage. Fig. 3.32 shows yet another arrangement.

I have also used the device with the motor on the bench at the end of the lathe. This involves long driving belts but was otherwise very satisfactory. In fact, with leather belts the additional length (and hence weight) of belting meant that the drive was quite adequate even when the travel of the saddle was considerable.

My final photo, Fig. 3.33, shows a similar arrangement as used on my little Boley watchmaker's lathe. The principle is identical, but the whole issue is made up from laboratory clamps and stands. The same stand holds my magnifying glass!

I hope this will give you some ideas. Simple? Yes. Cheap – yes again, for the motor and most of the rest came from 'Jack the Scrap' (and you need a motor whatever design of gear you use). Effective – indeed yes, and I wish I had set it up 40 years ago!

Fig. 3.31 *Standard clamped to the front of the cross-slide, Potts spindle vertical.*

Fig. 3.32 Standard clamped to the rear of the cross-slide.

Fig. 3.33 Spindle drive on a Boley watchmaker's lathe.

A truly mobile handrest

The use of a graver, toolmaker's scraper or even a proper hand tool for forming radii on corners or doming the ends of workpieces is common practice, but it has to be accepted that the handrest provided for most small machines leaves much to be desired. It is difficult to fit and awkward to use – often the saddle gets in the way. The 'Quick Fit' rest devised by Mr George Thomas (*Model Engineer*, 1997 p.157 et seq) is an improvement, but involves a major constructional project. Neither can be used comfortably on workpieces of any length (see Fig. 3.34).

Fig. 3.34 A memorial altar cross, about 16 inches tall, hand-turned in brass.

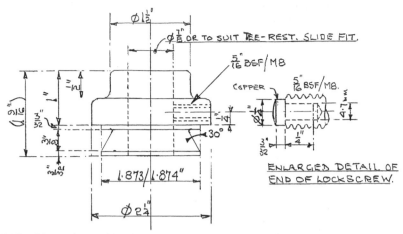

Fig. 3.35 Dimensions of 'mobile' handrest to suit the Myford Super-7.

The little device shown in Fig. 3.35 fits *on* the cross-slide and provides both convenience in use and the maximum choice of position. The drawing shows a design for the Myford Super-7, but a flanged type for use on other lathes would not be difficult to devise. Mine is made from aluminium alloy and after some 25 years of use is showing some few signs of distress. An iron casting would be preferable, or machine from $2\frac{1}{4}$ inch round cast iron stock.

It is a simple turning job, the only points to note being:

(a) that the upper edges should be well radiused as shown
(b) the bore should be dead square to the face of the $2\frac{1}{4}$ inch diameter flange
(c) the tool used to form the 30° taper should have a small (say $\frac{3}{64}$in.) radius nose.

Good finish is needed on this taper. The hole is, of course, bored to be a smooth fit to your tee-rest stem.

The dimensions shown should be adjusted to suit your machine so that the tee can be adjusted to bring the top of the thickest tool to centre height, in which connection I must emphasise that you need a **metal-cutting** tee with a flat top, highly polished (that normally supplied for woodturning is not suitable).

With this device the difficulties of forming shapely curves on wheel handles, clock arbors, and even ball ends, disappear as hand and eye work together – no more fiddling with topslide and cross-slide indexes and no more brutalising of workpieces with files and emery!

Hand-chasing screw threads
An unexpected bonus from this rest was the ease and speed with which screw threads could be cut. Anyone who has seen old craftsmen brass-finishers at work will recall how quickly they cut BSP threads. The problem for those of us less skilled lies in *striking* the initial thread. This can now be overcome.

90

Set up as for normal screwcutting with the machine in middle speed back gear, say 50 to 60 rpm. With the rest to the right of the work engage the half-nut. Hold the chaser firmly on the rest with one hand and advance it to cut with the other. It helps if the chaser is slightly angled sideways so that each tooth cuts slightly deeper. Make two passes (later, after practice, you may find that one is enough).

You have now a part-formed thread to guide the chaser. Disengage the leadscrew and throw out the back gear to run at approximately 25/35 ft/min. Advance the chaser to the formed grooves at the beginning of the thread and take a cut full length. Repeat this until you see that the crest of the threads is approaching completion, then try on the test nut. You can then 'shave' the thread as and where required to get a smooth fit throughout.

This does need a little practice but not as much as you might think. It is hardly worthwhile setting up for short threads, but for any longer than an inch the procedure certainly saves time. Fig. 3.36 shows a $\frac{1}{2}$-inch dia × 12 tpi thread in brass which took about 4 minutes from start to finish excluding preparatory work such as chucking, facing and centring the end etc. Of course, I did have the tee rest set up from a previous job – changing from topslide to rest and back again might have added a couple of minutes to the time.

Rigidity of lathe tools

It is a truism to say that model engineers use their lathes on work far larger than they are designed for. In the early days the normal machine would have been of 5 in. centre-height – true, it was treadle driven, but the bed and saddle were proportioned to the size of the work, not the driving power. The leadscrew of my daughter's $3\frac{1}{2}$ inch

Fig. 3.36   A hand-chased $\frac{1}{2}$ inch Whitworth thread.

Britannia was 1 inch diameter and the toolpost is designed for half-inch square tools. Indeed, the smallest size of tool listed in their catalogue of 1896 is $\frac{3}{8}$ in., intended for their light $3\frac{1}{2}$ inch machine. Modern lathes are, of course, much stiffer in their construction, the designers having made use of a better distribution of metal rather than sheer mass. However, when it comes to the tool in relation to the work we seem to have moved backwards. Many model engineers perhaps do not realise that the toolbits they use are really intended for application in a holder, the smallest of which are about $\frac{5}{8}$ in. deep, with only the actual point of the bit projecting.

Our tools are usually supported on a relatively weak pyramid of sliding components – there are five 'joints' between the lathe bed and the tool clamp, and even more with some tool support systems. Not the least of the advantages of the rear parting tool-holder is that most of these sliding supports are eliminated. This gives a pointer to the basis of the many problems met with by model engineers when dealing with machining operations such as parting off, using form-tools, and turning slender workpieces like crankshafts. Some of the devices used to ease other problems – the quick-change tool-holder, for example – may aggravate the difficulties of lack of rigidity. They are appropriate enough in the industrial setting, where even small lathes may weigh half a ton, but create difficulties in circumstances where tool forces are high, or long projection is needed. The two devices which follow were designed to overcome some of these problems.

Tangential tooling

The problems just mentioned are overcome in production workshops in a variety of ways, the chief of which is probably the tangential tool. In all cases where the feed is diametral – that is, plunging into the workpiece – the classical overhung cantilever type of tool is liable to chatter and 'dig in', especially on lathes as light as the majority used by model engineers. The tangential tool is virtually free from these problems, and is illustrated diagrammatically in Fig. 3.37.

It will be seen that the main cutting force (downwards in a plunging cut) is transmitted directly to the slides. No bending forces are involved in this plane, the only overhang being that due to the very small front clearance angle. Further, in the case of a form tool, resharpening does not reduce the resistance of the tool to the cutting forces, as only the top face of the tool is ground.

There is a small bending stress in the vertical plane, but this is very small, partly due to the deep tool section, and partly to the presence of the clamping device, not shown in the sketch. The result is that large and fast cuts may be taken with excellent finish and no risk

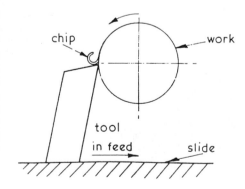

Fig. 3.37 *The principle of the tangential tool (tool clamps not shown).*

of disaster provided, of course, that the workpiece is strong enough to stand the strain; all too often in model engineering it isn't!

Naturally, there is a snag so far as we are concerned. Such a tool does need a sophisticated tool-holder, with provision for adjusting the tool height and for setting both the attitude and the angle of the cutting edge. I *have* built up a tangential parting-tool before now, but it was a tedious affair to set up and though I could part off up to 3 in. dia with it, it was, quite frankly, more trouble than its rare use warranted.

I have also, from time to time, supported wide form tools from the cross-slide with toolmaker's jacks. Again, very effective, but useless in many applications where the jack would foul the workpiece. It is important to appreciate that any such support *must* be secure; not only the work, but the machine as well may become more than a little disordered if the support comes adrift during the cut!

This is especially the case when turning crankpins, and it is just when machining this type of component that the limitations of the normal cantilever type of tool become apparent. And, of course, it is just in this situation that there is no room for a normal screw-jack. Very difficult – so all the more reason for doing something about it!

Fig. 3.38 shows one answer. The $\frac{1}{2}$ in. $\times \frac{3}{16}$ in. strip acts as packing beneath the tool proper, the end marked 'L' being reduced in width to correspond to the business end of the crankpin turning tool, and being the length necessary to clear the crankwebs (and balance weights if forged with the shaft). In fact, 'L' can be a bit less than this, as the work rotates about a centre some way above the top face of the device.

The vertical support is a piece of $\frac{1}{2}$ in. or $\frac{3}{4}$ in. wide strip, in the case shown $\frac{1}{8}$ in. thick, brazed to the underside of the flat; weld it if you like, better still. The height 'H' is made a few thou greater than the difference in height between the cross- and topslides. This extra ensures that the foot is truly taking the cutting load and not just floating; if it doesn't *quite* touch the cross-slide in service, chatter will be inevitable. True, you can put in a shim, but it is better to make it right in the first place.

The underside of the tool must, of

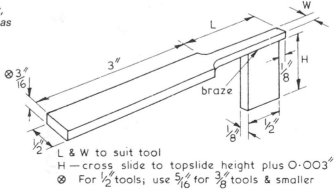

Fig. 3.38 Tool support, giving the same effect as a tangential tool.

L & W to suit tool
H —cross slide to topslide height plus 0·003″
⊗ For ½″ tools; use ⁵⁄₁₆″ for ³⁄₈″ tools & smaller

course, be fully in contact with the support, and my own practice is to grind the underside of the tool very slightly concave over its length. When clamped down, this straightens out and brings the underside of the nose of the tool hard down on the support. To correct for centre-height, packing is used between the tool and the support, not underneath the latter, for obvious reasons; and for equally obvious reasons, this packing must extend right to the nose, suitably reduced in width.

There are limitations. It can't be used with a normal four-tool turret, as there isn't enough depth, but the turret offers little advantage when crankpin turning and you shouldn't use tools of any great overhang in a turret anyway. Secondly, you can't traverse with the topslide, but must engage the lead-screw and use the calibrated hand-wheel instead.

A few minor points. It is fairly impor-tant that the support pillar is square to the flat, sideways. A little fore-and-aft lean doesn't matter, but if it leans side-ways it may throw the tool over on a heavy cut. I made mine $\frac{1}{2}$ in. wide as shown, but $\frac{3}{4}$ in. is better, as this then always spans the boring table slots by a fair margin. The length shown as 3 in. should be such that this part of the device spans the full width of the topslide, and any packing used should also run this full width. The thickness, shown on the sketch with an asterisk, can, with advantage, be greater; it is best arranged so that the tool point comes to centre-height with the mini-mum of packing. The stronger the thing is, of course, the better it will work. Similarly, if you are going mainly to support wide form tools, then the up-right part can be as thick as you like – even a round column.

The device shown in the photos (Figs 3.39 and 3.40) has been used to

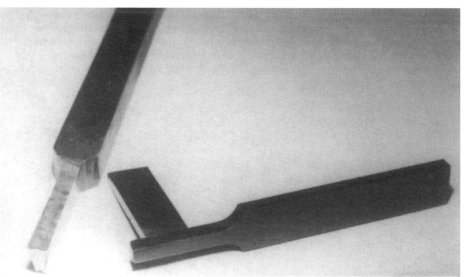

Fig. 3.39   A crankshaft turning tool and its support.

Fig. 3.40 *The tool and support set in place. It is now normally used in the Gibraltar toolpost (see next column).*

machine cranks of 2 in. stroke, with the tool point overhanging about $1\frac{1}{2}$ in. from the front edge of the slide. This tool by the way, is at least 70 years old, carbon steel, which I obtained at a sale a few years ago, and very useful it has been, too. It is really a $\frac{1}{4}$ in. wide parting tool, but I use it solely for crankshaft turning. The tool point is, of course, shaped as in Fig. 3.41, and is traversed from side to side while cutting.

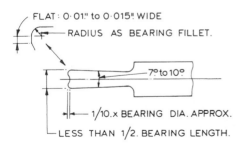

FLAT: 0·01." to 0·015." WIDE

RADIUS AS BEARING FILLET.

7° to 10°

1/10.x BEARING DIA. APPROX.

LESS THAN 1/2. BEARING LENGTH.

Fig. 3.41 *Tool planform used by the author for crankshaft machining.*

'Gibraltar' – a really rigid toolpost

Not too long after making the tangential type tooling described on page 92 I had an even more flexible shaft to machine, and as the topslide could not be used with the tool-point support there shown I decided to design a toolpost to sit directly on the saddle. There have, of course, been hundreds of different styles of tool-holder described over the years, all claiming some advantage or other. Most were devised either to ease the changing of tools or to facilitate adjustment to centre-height. My objective was different – to achieve maximum rigidity without sacrificing too much in other directions. The use of a tool-holder mounted on the cross-slide itself would do away with the flexibility of the topslide for one thing, and also permit a more rigid holder – one can't do much on the confined space available on the topslide. I also hoped that I should be able to devise something which would enable me to run close up to the tailstock; it has been a grumble of mine for decades that the designers of this component seem to have forgotten that it is no use having a strong support at that end if you can't get at the workpiece to machine it without having an excessive tool projection. I have already found that there were no problems when using the tangential-type tool-support arising from the absence of topslide movement; indeed in some ways the use of the leadscrew handwheel seemed to ease matters rather than the reverse.

The first, prototype, toolpost was made from a hefty chuck of DTD130B aluminium alloy and is shown in Fig. 3.42. It did everything I had hoped for and more. Fig. 3.43 shows it in use during a 'metal-shifting' session, with a cut of just over $\frac{3}{8}$ in. deep at $3\frac{1}{2}$ thou/rev

95

running at 70 ft/min. in En3. I was able to run right up to the tailstock with minimum tool projection and with the tailstock poppet within half an inch of the body. I was also able to use tools of larger cross-section than usual – the hefty $\frac{1}{2}$ inch butt-welded tools normally used on my hand planing machine. The tool gap does, in fact, permit the use of $\frac{5}{8}$ in. square tools if desired. There was a last-minute refinement, too; the little $\frac{3}{16}$ in. dia vertical peg seen in the photo, which carries my Verdict-type last word dial indicator!

On several normal jobs the facility with which the whole issue could be rotated was very helpful and, of course, in facing jobs the holder can be rotated so that the tool may be either at the front or the back when looking at it from the operator's position. For its initial purpose it worked admirably, as seen in the photograph (Fig. 3.44) and the long three-throw shaft was fully machined without incident. (A note about the

Fig. 3.42   The first Gibraltar toolpost made for trial.

Fig. 3.43   Shifting metal!

Fig. 3.44 The Gibraltar together with the tangential tool support makes for maximum rigidity at the tool point.

various clamps and thrust pieces used appears on page 56.)

There were some snags. These did not apply to the crankshaft, which was turned between centres and could be changed ends as needed, but as the tool was virtually at the centre of the topslide (on the Super-7) it was possible to get close to the chuck only if the swarf guard was taken off, and this was a nuisance. However, as the device had been so successful I arranged for castings to be made and altered the design to avoid this particular difficulty. This was shaped so that it could be used either on the Super-7 or on the older type ML7 machines – see Fig. 3.45 and the drawing Fig. 3.46. The casting is available from **Hemingway**, Wadworth

House, Greens Lane, Burstwick, Hull HU12 9EY and is in cast iron – better than light alloy for such a job, even if a bit heavier (which is not in itself a bad thing for a toolpost).

Grip in the four-jaw by the base and after setting true face the pads on the top. Then reverse in the chuck – you will see that I have provided a chucking-piece to ease this. I suggest that you leave this on afterwards; it does no harm and may come in handy if ever you want to alter things. Machine the base to suit whichever lathe you have, noting the little undercut adjacent to the spigot in both cases. If working to the ML7 drawing you will have rather a lot of metal to machine away, but the alternative of providing a different casting

Fig. 3.45 The production model of the Gibraltar. Compare with the final design. (Photo Neil Hemingway)

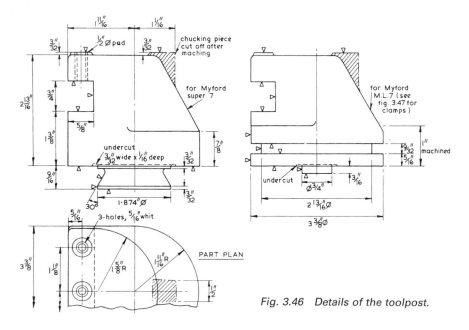

Fig. 3.46 Details of the toolpost.

would have added considerably to the cost.

I was able to machine out the slot on my miller, using an *Autolock Rippa* cutter, but I suggest you do it with normal slot drills; the lathe is *not* a milling machine and will not cope with very heavy cuts. Set the casting upside down on packing on the cross-slide and mill out a slot say $\frac{3}{8}$ in. wide to full depth. Then adjust the packing thickness to clean up to the full width, tackling first one side and then the other. I recommend that you make the tool slot a little over the $\frac{3}{4}$ in. shown. The $1\frac{3}{8}$ in. height will allow the use of $\frac{5}{8}$ in. tools, but if the point has been ground down you may not be able to bring it to centre-height otherwise. There is plenty of metal for the screws and the material is 17-ton high grade cast iron, though very easily machined.

When machining the base of the ML7 type I suggest that you use a carbide tool if you have one – not so that you can work faster, but because when chewing off the surplus metal in reducing the square corners to round you will, part of the time, be skimming the surface of the circular back of the casting. Aim to get a good finish on the bottom face of the circular groove which takes the clamping pieces (Fig. 3.47). These are made from mild steel, shaped as shown on the drawing, with a small bull-nosed packing riveted or brazed to the outer end. I have asked you to make four, but in many applications you will need only a pair, opposite each other. The holding bolt is a standard Myford tee-bolt shortened, but you may care to use tee-bars in the slot of the cross-slide and Allen screws with washers. (I don't like these screws here, by the way; the socket head gets full of swarf, and a small

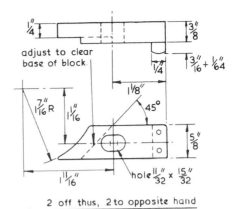

2 off thus, 2 to opposite hand

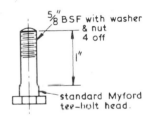

Fig. 3.47 Details of clamps and bolts for the ML7 version.

Allen key is always more likely to get lost than a spanner!)

Drill and tap for the tool securing screws, which can be hexagon, square or Allen heads to suit yourself. I have shown three such, as there is little point in having a rigid tool-holder if you don't hold the tool rigidly in it! The drawing does not show the hole for the dial indicator peg, but you can fit this in to suit your own type of indicator. This I would regard as a must – it saves removing the tool and can be set up and taken down so quickly that you will use it when you should and not just when you can find time to fit the thing up! I finished my own with a coat of primer, a couple of undercoats and a top coat of more or less Myford grey (I believe they are painted differently now) so that when in use it looks part of the machine (Fig. 3.48).

I don't claim that this accessory is the sole and definitive answer to the toolpost problem. It does not, for example, supplant my four-tool turret (which does, of course, remain on the topslide when using the Gibraltar) for there is no need for complete rigidity in much of my work. There is the added point that there are some jobs where the topslide is needed – in some cases it is essential, when turning short tapers, for example, or for 'picking up' a thread in screwcutting a job which is partially

machined. For short traverses not under power (and you should use auto traverse on all *but* short travels) it does take a little time to get used to using the leadscrew handwheel instead of the knob on the topslide. However, once you *are* used to it you will find, as I have done, that it is perhaps easier to work to a definite setting on the index. Those on the leadscrew handwheel seem to be easier to work to than the little ones on the topslide. However, despite these limitations the advantages far outweigh them. For heavy roughing cuts it is a wonder, and on long finishing work I have found that somehow the tool seems to stay sharp much longer. It is virtually essential when carrying out any metal-spinning; see Fig. 3.49. For

Fig. 3.48   Even with very short tool projection work can be done right up to the tailstock.

Fig. 3.49   *Using the Gibraltar when spinning the domed top for a boiler (Photo Mike Chrisp)*

the purpose for which it was first devised – machining crankshafts – it is *far* superior to anything else I have used except, perhaps, the toolpost on an 18-inch crank-turning lathe I used over half a century ago! This, of course, had no topslide.

Taking very fine cuts – 'shaving'

This is a technique rather than a device, to meet the situation when the odd few tenths of thousandths of an inch must be removed from the workpiece. Many will resort to emery cloth – this not only risks tapering but is certain to round any sharp corners. However, given a

really sharp tool it is possible to take very fine cuts indeed.

It is vital that the tool be honed razor sharp and set at centre-height – definitely *not* above. The front clearance should be about 7° and the top rake 12° for steels, 6° for cast iron and zero or minus 2° for brass. To avoid waviness the front planform should be a radius of not less than $\frac{1}{8}$ inch (3 mm). Power feed is essential, at about 0.004 inch (0.1 mm) rev. The cutting speed should be about 10 per cent lower than 'normal'.

To put on the cut an experienced turner would simply tap the cross-slide handle very lightly. Such skill is not given to all of us, and we must rely on the machine instead. This is done by setting over the topslide at an angle, so that its index divisions are reduced by a known factor. At 5°44′ this factor is 1/10, but is difficult to set. However, at **six** degrees an index movement of 0.001 inch becomes a depth of cut of 0.0001048 inch which is near enough for anyone; if attempting a cut of one-tenth of a thou we are not going to worry about the error of 4.8 millionths!

In practice it is best to set on cuts of not less than 0.0002 inch (0.005 mm) to give the tool edge a reasonable amount of work to do. It is also advisable to tighten up the cross-slide gibs a trifle (or even lock it) to avoid accidental movement here.

This technique can be used when boring provided that the tool is reasonably stiff, but it is advisable to take the cut in the reverse direction – from the headstock outwards.

To reiterate – the tool *must* be honed, preferably with hard Arkansas stone, to a razor sharp edge, finishing the top surface last. A smooth surface here is as important as the sharpness of the edge.

Heavy drilling in the lathe

A lathe fitted with backgear can be used to drill quite large holes if the drill has a suitable Morse taper shank to fit the tailstock. I must confess that I go quite slowly and gingerly as there is always a risk of the drill 'grabbing' and turning in the taper. On a normal drilling machine the tang on the drill will engage and prevent this, but there is no corresponding slot on lathe sockets. Yes, I know that the 'drive' should be taken by the taper, and so it is; but the presence of a tang does prevent rotation should the grip on the taper be lost. However, it is

sometimes necessary to take the risk, and I have used up to $\frac{7}{8}$ in. drills on the Myford.

Occasionally there is a problem; the workpiece cannot be held either in a chuck or on the faceplate and must be carried on the saddle, with the drill in the headstock mandrel. The risk of slippage is still there but as most headstock sockets are hardened these days it isn't quite so great. There is, however, another hazard. The normal method, mentioned by many writers, is to use the saddle handwheel or, as I do for smaller drills, power traverse. With large drills this will put an excessive twisting force on the saddle, increasing the friction losses, and overloading either the rack pinions or the leadscrew half-nut as the case may be. The forces in drilling are very large – on a $\frac{1}{2}$ inch drill in steel, about 150 lbf. If you don't believe me, set the bathroom scales on the drilling machine, a vice on the scales, and drill a hole. You'll see!

To get over this, why not use the tailstock to push the saddle along? Fig. 3.50 shows a casting being drilled, with two bosses requiring the holes to be parallel and in line. It has been clamped to the saddle on packing and a $\frac{1}{2}$ in. drill is in use. There is a piece of dowelling between the tailstock poppet end and the work, so that when the drill breaks through this takes the cut (it starts to rotate as a rule, but this makes no odds). All the thrust is now contained in a straight line and no damage can be done anywhere. Further, all the strain is taken off the clamping screws. Note, by the way, the spare backplate on the mandrel nose. This is always fitted whenever the thread will be exposed in service, to avoid risk of damage. You will see that it is painted white; this reflects light onto the work when the

Fig. 3.50 *Using the tailstock to apply thrust when drilling from the headstock.*

saddle is so close to the headstock that the machine lighting cannot reach. Note also the turned cover in the hole which normally is filled by the topslide, to prevent swarf getting onto the cross-slide feedscrew.

Cutting fluids and accessories

Cutting oils
I do not intend to discuss the relative merits of soluble and straight cutting oils. I use both – for general service, a relatively light straight oil (enriched with an additive), pump circulated and piped to serve either the lathe or the milling machine. For more difficult subjects I use soluble oil. Soluble would, of course, serve for all, but as there can be long intervals when the workshop is idle I prefer to avoid the need for a complete clean-up each time the machines are used.

Most of you will be aware of the corrosion which can arise from soluble oil, which is, more properly, an *emulsion* and not a solution. This corrosion is not rust caused by the water, as many believe, but a rather special form of attack brought about by anaerobic bacteria present in the fluid. In the presence of oxygen they do no harm, but if air is excluded these bacteria multiply, feeding on the emulsion and releasing corrosion products. The commonly suggested expedient of using a higher proportion of oil (and hence, less water) is the *wrong* thing to do – all you are doing is providing more 'fuel' for the bacteria! The recommended mix of

20/1 to 30/1 is quite rich enough.

This causes little trouble on industrial machines, as they are in continuous use and are regularly and thoroughly cleaned down. But it can cause problems for those of us who are able to get into the workshop only for odd hours at a time.

There is not much risk with the lathe slides, as only a small amount of fluid is trapped, and little air can get in. If the oil-gun is applied after each campaign on the machine this should drive out what soluble oil has got there. However, it has to be accepted that when felt wipers are fitted these do provide a truly gourmet restaurant for bacteria! My own drill is to remove these altogether when using emulsion, and replace them (after cleaning and drying) when changing back to straight oil. Whenever a long period of cast iron machining is expected I take off the oily felts and replace with dry. (These felts

are very cheap and it pays to have spares available.)

Cutting oil feed

As mentioned earlier, my normal supply is from a pressure pump mounted in a 2-gallon tank, which is arranged with a 2-way cock to feed either the lathe or the miller, each with a drain pipe back to the tank. The drip tray of each is arranged with a 'fall' of about 1/60 towards the drain, with a strainer to keep out swarf. That on the lathe can be plugged when using soluble oil. Alternatively if a long campaign on this fluid is expected, a separate drain pipe and receiver can be fitted.

Soluble oil is dispensed from the can shown in Fig. 3.51. This was rigged up nearly 50 years ago now (following an article by the great Edgar Westbury if my memory serves me correctly) and can be swapped for the pump-supplied standard in a few seconds. It is simply a

Fig. 3.51   The drip can used when cutting oil other than that in the pump system is needed.

pint can to which a 20 swg tinplate bracket is soldered, the bracket having a piece of brass tube between the folded-over ends. The upright is $\frac{1}{4}$ inch diameter steel, with, as you can see, a nut and soft spring washer at the top; just tight enough so that it stays put but is easily moved. The base arm is $\frac{1}{2} \times \frac{1}{8}$ inch steel, held by a tee-bolt to the cross-slide.

The outlet is an old swivel gas tap, but a normal petrol cock with the now well-known *Locline* tube system would, I'm sure, be more useful. An important point is that the end of the tube must either be cut at a fine angle, or a nozzle used, so that the flow can be concentrated, even when a mere drip. This simple device served me well for many years, and can be commended to those who cannot afford – or do not have the room for – a pump and tank system.

Splash guards

As soon as any form of drip or flow feed is used the waste thrown off the rotating chuck must be confined, for obvious reasons! Even with the drip can in Fig. 3.51 this can be a nuisance – as I found when I built my 'new' workshop back in 1972. The nicely painted wall behind the machine soon suffered. The solution is shown in Figs 3.52 and 3.53 which are, I believe, self-explanatory. The guard is attached to the headstock using the bolt-hole provided for this purpose on the machine. It can be swivelled up and down at will. I would recommend, however, that the 3-inch dimension be made slightly longer to clear the rear of the cross-slide when working close to the chuck. The bracket is $\frac{5}{8} \times \frac{1}{16}$ inch steel and the sheet 18 swg aluminium, painted to match the lathe.

An **extension to the drip tray** was made later, shown in Fig. 3.54. This was needed initially to keep fast-flying swarf

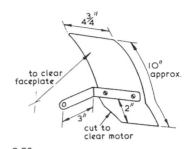

Fig. 3.52

CHUCK SPLASHGUARD

Fig. 3.53   Rear splash guard for the chuck.

(chiefly brass) off the shelf behind the machine on which I keep chucks, centres etc. – a foolish arrangement really. But it was there, built into the general set-up round the machine, and this guard served well enough. Later, when I fitted the pump system and was able

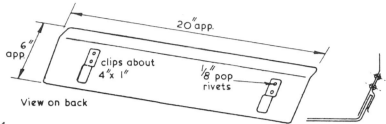

Fig. 3.54

to use a full flood of coolant to the tools, it became even more useful.

Extended chuck guard
This became essential when the pump system was fitted. It is shown in Figs 3.55 and 3.56. The base has four little magnets (available very cheaply nowadays from surplus stores) which secure it to the drip tray. The material used for the guard itself was $\frac{1}{8}$ inch perspex, easily bent if gently warmed, the idea being that I could see through it. In the event, after about ten years, it has now relapsed into translucency, and aluminium would have done just as well. The top end is secured with a spring (bulldog) paperclip, bolted to the top of the rear guard. It is a simple matter to

Fig. 3.55 Detachable front chuck guard. The hinged section is secured by a bulldog paperclip attached to the top of the rear guard.

Fig. 3.56 The guard open to give access to the chuck key.

release and swing down the front guard to use the chuck key.

Milling machine vice tray

Like most pedestal machines, the Myford VMC has a coolant drain to the table, another on the base of the column and a third to the sub-base collecting tray into which the others drain. The great advantage of pump feed is, of course, that the oil flow can be jetted to remove chips as well as to lubricate the cutting edges; very important when slotting, especially with small cutters.

Unfortunately, despite the comprehensive collecting and drainage system, much fluid falls to the floor simply because the machine vice projects well outside the sides of the table. To cope with this I made up the vice tray seen in Fig. 3.57. This is simply a cook's shallow baking tray, about $10 \times 12 \times \frac{1}{2}$ inch deep – as used for making fudge in this household!

Holes are made to pass the vice h/d bolts, the vice register gluts, with two further holes to line up with one of the tee-slots at the front of the table. In addition, a short stub of $\frac{3}{8}$ inch brass tube is soldered into the rear corner, from which a hose runs to the sub-base tray, terminating adjacent to the drain. There is a small loose cover over the entry to the stub to prevent swarf from falling into the tube directly.

This device reduces splash problems to negligible proportions, though it is necessary to clear the tray of swarf fairly frequently, as this can impede free drainage.

Recovery

A considerable amount of fluid is lost when disposing of swarf. This is hardly worth bothering about with soluble oil, but neat oil is relatively costly. Try this. Make a fairly large number of small holes in the bottom of a tin; I use the

Fig. 3.57 Tray to collect drips from the machine vice.

large *Nescafé* tins used by bulk consumers (of which I am one!). It needs to be at least 6 inches in diameter and 8 inches or more deep. A wall-paint can would serve just as well. Shovel the oily swarf into this, and stand it on top of a similar (but 'unholy') tin. Leave for several days after which almost all the oil will have filtered through. This can then be tipped back into the main tank.

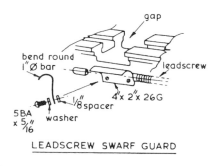

LEADSCREW SWARF GUARD

Fig. 3.58

Leadscrew guard

For some obscure reason the designers of lathes always provide a leadscrew with a thread extended well beyond the reach of the leadscrew half-nuts – at both ends. I don't know why! The last few inches at the headstock end are redundant, but they lie just where the swarf falls onto the screw. Yes, I know there is a little guard on the saddle, but it doesn't cover the screw except when close to the headstock. Now, it might be argued that dirt on a bit of screw that is never going to be used does not matter. But apart from looking bad there *is* a chance of such dirt being carried back under the saddle apron. Indeed, the maker's little guard does just that when any bits of long curly swarf get onto the leadscrew.

Hence Fig. 3.59. It is made of 26 swg sheet aluminium, and held by two 5 BA (or $\frac{1}{8}$ Whit.) screws and washers. Note that it goes *behind* the leadscrew, not in front as the sketch (Fig. 3.58) may

Fig. 3.59   The leadscrew swarf guard fitted.

suggest. It is so arranged that when the saddle is at its extreme left-hand position the little guard attached to the saddle passes over it without fouling. You will note that it is packed away from the lathe bed so that any coolant running down can get away without being trapped.

My reminder

Guarding against dirt is one thing, but human error is a more serious matter, like engaging the leadscrew half-nut for a feed of 6 thou/rev only to find that you are about to cut an 18 tpi thread instead. This can be quite an experience at 1800 rpm with the saddle belting along at $1\frac{1}{2}$ inches/second! So I made the device shown in Figs. 3.60 and 3.61 just to remind me where things were set. The little label has 'Fine Feed' painted on one side and 'Screwcut' on the other. There is a further label under this one, seldom used, which reads 'Special' which is for use when I have the auxiliary quadrant set up for metric or other special threads (my lathe has a screwcutting gearbox). You will notice on the sketch a couple of arrows painted on the casting alongside the tumbler reverse lever. This indicates the direction of travel of the saddle at

Fig. 3.61   Fine feed and reverse indicators.

the two settings. This is rather more important than usual in my case as I do a fair bit of work on the Lorch screwcutting lathe and on this the reverse lever works the opposite way to that on the Myford (with the normal screwcutting train on the banjo).*

Naturally, I have to remember to alter the labels! In fact, they remind me to, for when the gearguard is opened they rattle a bit against each other!

Protecting the taper sockets of the lathe

The biggest enemy of taper sockets is dirt, for the presence of even a tiny particle can cause the centre to run badly out. Worse, any such in the tailstock while using a drill chuck can cause the arbor to spin and this in turn will groove the socket, raise a burr and cause endless trouble in the future. There are many ways of cleaning – a piece of rag on a wire can be used as a 'pull-through' in the headstock – but many tailstocks these days have no through hole (more's the pity) and this

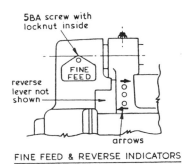

FINE FEED & REVERSE INDICATORS

Fig. 3.60

*My Super-7 has a gearbox.

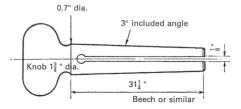

0.7" dia.

3° included angle

Knob 1¾" dia.

31¼"

Beech or similar

Fig. 3.62   Centre cleaning tool. Also serves as a stopper when no centre is in use.

cannot be used. Fig. 3.62 shows the gadget I made, turned up out of hardwood. I used boxwood but any close-grained wood will serve. The length should be longer than the length of the taper otherwise when fitted with a square of cloth it will not reach the bottom end. I use 'four-by-two' flannelette familiar to many for using to clean out rifle barrels and obtainable from gunsmiths. A piece is threaded into the slot and the whole given a couple of twists in the socket. Repeat if the first application brings out a lot of dirt. The plug, without the cleaning cloth, sits in

my tailstock barrel when I am not using it. It keeps out dirt and saves catching the back of the hand on a centre or chuck while using the lathe.

The same device is, of course, used to clean out the headstock socket, but here there is another problem. There is risk of swarf getting right down the hollow mandrel, particularly when drilling. I use a liner, as shown in Fig. 3.63. The taper head is first bored and then turned to No. 2 MT, the bore to fit any convenient size of tube you have available. The length of the wooden taper need be no more than an inch or so. My own tube will pass up to $\frac{7}{16}$ in. diameter so that it can remain in place for a great number of machining operations. The wooden ring at the back end is to prevent the tube from rattling about, and the tube is made of sufficient length to project outside the change-wheel cover, but not so long as to prevent the cover from being opened. Any swarf which does pass through will not then land amongst the headstock gearing.

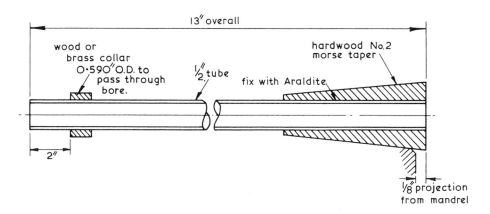

13" overall

wood or brass collar 0·590"O.D. to pass through bore.

½" tube

fix with Araldite

hardwood No.2 morse taper

2"

⅛" projection from mandrel

Fig. 3.63   Headstock bore protector. The dimensions are those for the Super 7 lathe.

110

SECTION 4

Miscellaneous

These are bits and pieces which could not be classified in any of the other groups. The section could have been very much longer as most of the home-made devices in my workshop seem to fall into this category. However, as there is a limit to the size of any book I have selected just a few which have, in the past, been considered worthy of publication.

Milling machine spindle speeds

The popular mid-range VMC-type milling machines offer a reasonable range of speeds (160 to 2540 rpm) but the belt system (Fig. 4.1) is such that there is a gap in the middle of this range. The speeds available are shown in the table, together with the nearest cutter diameter appropriate to mild steel (for high performance cutters).

Unfortunately it is not possible to set a belt directly between the motor and the existing spindle pulleys as the jockey pulley arm gets in the way. However, if

a further pulley can be added *above* that on the motor the problem can be solved – see Fig. 4.2. This belt just embraces the jockey which acts as a vibration damper.

The motor pulley boss is longer than the length of the motor shaft, so that it is possible to fit a stub shaft within, having an extension to fit the new pulley (see Fig. 4.3). The keyway is made to suit the existing 3-step pulley. (Note that the dimensions here will depend on the motor used.)

The new pulley may have to be machined on the face of the boss to ensure that the head of the new (longer) retaining bolt clears the gear cover. Fig. 4.4 shows the assembly.

This alteration provides a speed of about 780 rpm, virtually in the middle of the gap shown in the table. The time taken was only the odd hour or so – followed by some days awaiting delivery of a new A55 belt (the exact size will depend on the pulley used) to replace

Belt Set	D-1	C-1	D-2	A-1	—	B-2	C-3	A-2	B-3	A-3
Speed, rpm	160	260	390	560	—	1090	1190	1350	2080	2540
Dia., inch	$2\frac{1}{2}$	$1\frac{1}{2}$	1	$\frac{3}{4}$	—	$\frac{3}{8}$	$\frac{11}{32}$	$\frac{9}{32}$	$\frac{3}{16}$	$\frac{5}{32}$

The very commonly used $\frac{1}{2}$ inch cutter falls into the gap indicated.

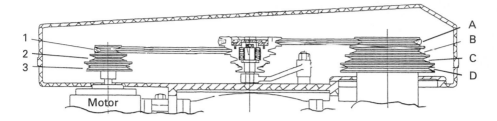

Fig. 4.1 Arrangement of speed pulleys on milling machine.

Fig. 4.2 Extra pulley and belt arranged to clear jockey pulley arm.

the ancient and much frayed relic found in my stores!

Filing buttons

It is not surprising that many engine or machine components have rounded ends, for many generations of fitters and operators have cut themselves on sharp corners, and 'design' is as much the result of experience as of analysis. The ways of making such rounded ends is legion, ranging from setting up complex milling rigs to the scribing of a circle followed by tedious 'filing to the

Fig. 4.3 Extension stub to extra pulley, sized to fit the motor shaft.

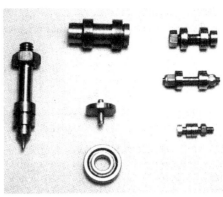

Fig. 4.5 A selection of filing buttons (rollers).

Fig. 4.4 The extra pulley fitted to the motor shaft.

Fig. 4.6 Filing rollers in use forming a forked shaft end.

line'. The filing button (or roller) is, perhaps, the simplest and (it would appear) the least known.

Fig. 4.5 shows a few from my collection and illustrates the construction well enough without the need for a drawing. A pin or screw is made a good fit to the hole in the lever or rod, and carries two discs or bushes, o/d to suit the desired radius, having a hole a close-running fit to the pin. Also shown in the photo is one of a pair of small worn out ballraces which make excellent filing buttons when of the right size!

In use, the two discs are set either side of the workpiece with a cotter. The work is set in the vice (Fig. 4.6) and the file applied so that when down to size it will roll on the discs without cutting. Initial cuts may be with a very coarse file, finishing being done with a Swiss file or even emery.

For one-off jobs it is scarcely necessary to harden the discs, but as they may come in useful again it is always worth doing so. There is no need to temper. I tend to use casehardened mild steel as it is cheaper than silver steel.

Making hollow mills or rosebits

There are occasions when a quantity of workpieces need the ends bringing down to BA (or other) sizes and it can be a tedious job turning them, even if the machine is fitted with saddle and cross-slide stops. Watchmakers use rosebits to machine the pivots on the end of their spindles, and there is no reason why larger ones should not be used in model engineering. In fact, they are available commercially, but at a formidable price! They are very handy indeed, as a single cut reduces the work to the exact dimension required; a

simple stop, or just the indexes on the tailstock poppet, will look after the length of the stub, and provided they are not overloaded by too fast a feed the finish is good. So, although I have some for my Lorch lathe – all far too small for model work – I decided to make a set to suit the Myford and to cover the normal BA sizes I use.

Fig. 4.7 shows the general dimensions. All save the largest are of $\frac{5}{16}$ in. silver steel, that for 2 BA being made $\frac{3}{8}$ in. o/d. The socket for the tailstock was made from a drill-chuck arbor; it could have been made from scratch, but this one was to hand and they are made of a very easy machining steel; besides which the taper hole in the end need not prevent it being used *for* a drill chuck if ever the need arose. The taper on the end of the cutters is quite arbitrary – you can make it what you

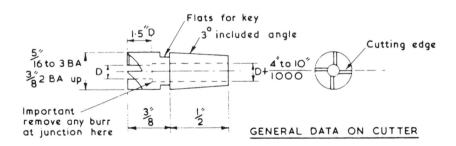

GENERAL DATA ON CUTTER

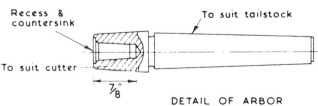

DETAIL OF ARBOR

Fig. 4.7

please – but that shown (3° included angle) is very nearly a Morse angle so that we can be sure it will grip well enough.

The first step is to make the taper hole in the arbor (and one in a fixture to be mentioned later) of such a diameter that the cutter fits without going in too far or not far enough. We must first make a test gauge. Chuck a piece of $\frac{5}{16}$ in. stuff (steel or brass) in the three-jaw, using paper if need be to get it true (if badly out you may have to use the four-jaw, but this is tedious, and an error of $\frac{1}{2}$ to $1\frac{1}{2}$ thou won't matter much). Set over the topslide by $1\frac{1}{2}°$ inwards towards the chuck, and turn a taper on the stock. Part off so that this taper is $\frac{1}{2}$ in. long, reverse in the chuck and put a slight bevel on the small end. Remove the chuck and set the arbor in the headstock making sure all is clean. Measure the small end of the little taper plug and centre and drill the arbor with a drill a few thou larger – I used $\frac{17}{64}$ in. Go about $\frac{7}{8}$ in. deep and keep the drill clear and the feed steady to avoid wandering. Now set up a tiny boring tool and bore the taper with the top-slide until the test plug taper sticks out only about $\frac{1}{64}$ in. Lightly bevel the edge of the hole and machine a recess about 10 thou deep as well, to protect the edge of the taper hole from damage, as shown in Fig. 4.7.

Remove the arbor and set up the four-jaw, to make the jig we shall need to mill the teeth. This is simply a *square* arbor which can be held in the vice on the vertical slide. Cut off a piece of $\frac{1}{2}$ in. square steel about a couple of inches long, clean up the exterior, and set it true to the faces in the chuck. Face the end, centre and drill $\frac{17}{64}$ in. (or as required) as before, and again machine the taper to suit the plug gauge. Mark

the jig for what it is, so that you can use it again if need be in the future. While you are at it, if you are likely to use rosebits in any other machine which has a different Morse taper, make an arbor for that also – one with a parallel shank is useful, as it can be held in any chuck.

Now for the cutters. The procedure is the same for all. Chuck the silver steel with about $\frac{1}{4}$ in. protruding, centre deep enough to accept the end of a drill of the appropriate size (see later), draw the work out until 1 in. protrudes, drill to size about $\frac{15}{16}$ in. deep. Turn the taper with the slide-rest, and finally part off to length. In the drilling, it is vital that the drill be kept free of chips so that it runs true. For the 2 BA chap, turn down the length behind the head to $\frac{5}{16}$ in. dia before making the taper. The cutters are then held in the chuck by the head taper outwards, and a slight bevel put on. The hole is then opened out with a drill about 10 thou larger than the one previously used, to the depth shown in the drawing (Fig. 4.7). Mark each blank with its BA size.

The drill sizes to be used must be chosen with some care; nearly all will drill larger than the nominal diameter and only one hole (that for 5 BA) suits a standard reamer. In most cases I put through two drills, as listed below, but *please* note that yours may be different. The trick is to drill a series of experimental holes, and check these against a well-made screw of the size it is supposed to suit. It is better to be a couple of thou small than even 1 thou large, especially if your dies are a bit tired. Here are the sizes I used, in mm.

2 BA No. 4.6 followed by No. 4.7
3 BA No. 4.0 followed by 4.1
4 BA No. 3.5 followed by No. 3.6

5 BA No. 3.0 followed by $\frac{1}{8}$in. reamer
6 BA No. 2.7 followed by No. 2.8

If using metric drills choose those which are nearest below these sizes.

After all the holes are drilled, the arbor is substituted for the chuck, and each cutter blank in turn is mounted therein and the end faced to a fine finish. The edge of the hole must be dead razor sharp, with no burr and no bevel, so hone the tool to razor sharpness, and take light cuts. Take off any burr at the junction of the two bores in the cutter blank.

The teeth may now be milled. I have given no detail dimensions, as these will depend on the diameter of the endmill or slot drill available; choose the sharpest you have, so long as it is of reasonable size. The object of the exercise is to get the flat face as nearly as possible on the centreline of the cutter, and the sloping face (formed by the diameter of the cutter) meeting the flat end of the blank at about 60 degrees. This last is not critical, but should not be less than 45 degrees. The next requirement is to get the teeth about $\frac{3}{64}$in. wide, measured on the flat face. Again, this is not critical – between 40 and 50 thou will serve. Finally, it is essential that when making the flat face of the tooth, the cutter should break into the central hole. This will present no problems down to 6 BA, but smaller than that may need the tooth thickness to be reduced.

Fig. 4.8 shows the required position of the endmill – I used one $\frac{5}{8}$ in. diameter. The vertical slide should be set up facing the chuck, and checked that the fixed jaw is parallel with the machine bed. The endmill is, of course, held in the three-jaw and again it will pay to use cigarette paper (etc.) packing

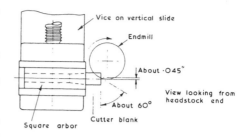

Fig. 4.8   Milling the teeth. See also Fig. 4.9.

to get it running true, if the chuck is at all old. One of the blanks (the smallest hole) is fitted to the square arbor and tapped in with a soft hammer, and the arbor mounted in the vertical slide vice so that the end is just flush with the side of the vice jaws. A stop will be needed to limit the saddle travel, and I used a toolmaker's clamp on one of the lathe shears. Set this first by eye so that the face of the endmill will be on the centreline of the cutter in the vice. Note that all cutting must be done with the face of the endmill, feeding in with the leadscrew handwheel, and not by traversing the cross-slide; once the correct setting is found, the latter should be locked. Similarly the vertical slide, using a clamp if need be.

Adjust the vertical and cross-slides and make a trial cut; run the cutter at about 40 ft/min. surface speed, and use oil to get a good finish. Adjust the cross and vertical-slides as necessary to approach the conditions listed above. Once you are satisfied that all is in order, lock up the slides and the saddle stop. Take a last cut in case anything has shifted, after which retract the saddle, release the vice, rotate the square arbor one flat, taking care that the end face of the arbor is again flush with the side of the vice, tighten up, and feed in with the leadscrew handwheel until you reach the stop. Allow the cutter to dwell for a few revs when you reach the stop. Carry on like this, until all four teeth are cut. The cutter can then be removed – you will need a drift to knock it out of the arbor – and should be examined under a glass for any imperfections. If all is in order, you may now deal with all the other blanks one after the other, except any which has a larger diameter head. For this one you will have to traverse the cross-slide a bit further back and experiment with the first tooth as you did before. You will probably have to adjust the vertical slide as well (Fig. 4.9).

Before leaving this part, there is one important point to observe. You must arrange the position of the slide relative to the cutter so that the latter is tending to drive the blank into the arbor, and so that the teeth of the endmill walk into the flat end face of the blank. This means, in effect, that the vertical slide

will be at the back of the cross-slide and the centreline of the blank will be below the centreline of the endmill. If, therefore, you use a milling machine instead of the vertical slide set-up, the same geometry must be observed. If otherwise then (a) the blank may well jump out of the arbor and (b) you will get a nasty feather edge on the business end of the cutter.

While the rig is set up, you should consider the means you intend to adopt to remove the finished cutters from the tailstock arbor. There are three common ways: (1) drill a hole right down the arbor and drive the cutter out with a drift, (2) drill a hole across the arbor, just about where the tail of the cutter will be, and use a taper drift to eject the cutter, in the same way that drills are shifted from the larger drilling machine and (3) machine a couple of flats on the cutter, for use in the same way that centres were removed on older pattern

Fig. 4.9 Cutting the teeth with the blank held in the square arbor.

centre-lathes. I used the last method, but either of the others would do. If you are going to mill flats, then you can do it now. Set up a $\frac{1}{8}$ in. or $\frac{5}{32}$ in. endmill in the chuck, mill one flat right across, with the endmill just clearing the end of the square arbor, and then rotate the arbor two flats and mill another (using the vertical slide to traverse, of course). Again, identical settings may be used for the cutters of the same diameter, but a different one for the largest. Make them fit any convenient BA size spanner, or make a key to fit.

The rest of the work is going to need a good light, a fair bit of care, and a very fine file – say a No. 4 or 6 cut precision Swiss file. If you haven't one of these, use a fine India or aloxite slipstone with thin oil, not one of the blue-grey slipstones. The advantage of the Swiss file is that it has a safe edge, but the stone will cut just as well; you must just take more care. Mount the cutter in the square arbor again, and carefully back off the face of each tooth at about 25 degrees, leaving a witness just at the front edge – Fig. 4.10a. This should be about $\frac{1}{64}$ in. wide; the exact width isn't critical, but get them all the same by eye. Take great care not to rub the adjacent flat cutting face, and still more care not to rub off a corner. If by some mischance you do take a corner off, mount the nearly finished blank in the No. 2 MT arbor in the lathe and skim the tops of the teeth again – very little cuts and a very sharp tool, please! You will find you soon get the hang of this job, and the square arbor simplifies matters, as you have only to rotate this in your vice to present each tooth to the file in the same attitude as the last. Finally, very cautiously, stone off all burrs.

The cutters may now be hardened. You will probably have your own ideas

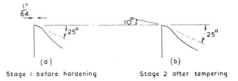

Fig. 4.10 Forming cutting edge.

about this, but it is important to heat the work up slowly, hold the heat at the right temperature for a while, and quench business end first. I arranged the cutters in a row, one behind the other, so that the waste heat from the first was pre-heating the others and so on. Look down the hole to see that you are at the right temperature and then hold for at least 60 seconds on these little chaps before quenching. The cutters stood on their teeth, on a piece of insulating brick (*Folsain, Fossalsil* or similar) and were gripped by the tapers for quenching. Remember to dry the tongs in the flame before using them again.

To remove scale I use a 4 in. soft brass wire brush at about 2,000 revs in the drilling machine. This put enough polish to be able to see the tempering colours or polish with a fine emery, but remember the cutters are now very brittle. Temper to pale straw. The pukka way of doing this is in a sand bath, but I simply held each cutter in the tongs at the end of the taper where it merges with the head and played a soft flame on the smaller end. Take care that the flame doesn't shoot through the hole or you will overdo it. Incidentally, I find that I tend to over-temper if I do this job in electric light and have either to wait until daylight, or make allowances. Alternatively, they may be 'cooked' in a deep-fryer for 20 minutes – a very effective expedient!

Clean and dry the cutters, and set up the square arbor in the vice. You can do with a little eyeglass for the next job too. This is to stone the cutting edge, at about 10 to 15 degrees just taking out the witness you left before (Fig. 4.10b). This needs great care and a good light. It is important to see that the stoned part is even, right across the tooth. Do this for all teeth on one cutter, using a fine India stone and oil, and then go round all the teeth again, very lightly, with an Arkansas stone at the same angle. If you haven't such a stone, then the finish left by a fine India will do, but the Arkansas gives just that extra bit of Rolls-Royce finish that pays dividends on the workpiece. The final step – with the cutter still in the arbor, but hand-held this time, is lightly to stone the flat front face of each tooth. Examine all under the glass, and retouch any deficiency. If you see any slight burrs on the outer edges, lightly stone these too. If by any mischance there is a burr inside the hole, and you have no stone small enough to get inside, chuck a piece of hardwood (pegwood as used by clock-makers will serve best of all) and run this down with the cutter in the tail-stock, about $\frac{3}{4}$ in. long. Impregnate this run-down length with fine emery-paste, and try to lap out the bore, but don't let the cutting edges come near the shoulder on the wood lap. The great risk with this process is that you may make the hole taper, so that when in service the machined peg will bind in the hole as the cutter advances. But if you have been careful in the earlier stages to remove any burrs after drilling and milling, no trouble should arise now.

In service, the cutter is mounted in the arbor in the tailstock, and run up to the work. It will cut surprisingly fast (it is the equivalent of four knife-tools working at once) but don't be tempted to force the pace or you will get a poor surface finish on the peg. Use plenty of cutting oil, enough to wash the chips away as well as to lubricate and cool the tool. Remember it is carbon steel, and adjust the cutting speed accordingly, to suit the outside of the work diameter bearing in mind that the tool is making a relatively wide cut on four faces at once (it is the heat produced by the cut that does the damage, as much as the absolute speed). I work mine at 40 ft/min. on soft mild steel, 35 or even 30 on BDMS, which is work-toughened, 30 on free-cutting brass, and about half that on drawn gunmetals. But the little fellows supplied with the watchmaker's lathes will cut very hard steel if the speed is kept down and plenty of oil used. The one material I don't use these cutters on, ever, is work-hardening stainless steel, but then, I try not to use this material for anything if I can avoid it.

Fig. 4.11 The finished article. Top: No. 2 MT arbor and cutter. Left: two cutters. Centre: the square arbor. Centre left: arbor for use in the Lorch lathe. Bottom: test piece run down with each cutter in turn.

There it is, then. You should now make a trial cut on a piece of (say) $\frac{1}{4}$ in. material to check the diameters of the finished jobs (see Fig. 4.11). Mine all came between 1 and 2 thou small except for the 4 BA cutter, which was 1 thou over-size. This was the only one, too, which showed any tendency to be tight on the work, suggesting a taper hole, so that it looks as if I need a new No. 28 drill!

A holder for throwaway endmills

Many model engineers will be familiar with the Clarkson range of throwaway endmills, but for the benefit of those who are not, I should explain that these are supplied at such a low price that the makers claim that 'resharpening is a nonsense'. I wouldn't altogether agree with that so far as we are concerned, but it is probably true that it would cost a *commercial* firm more to sharpen one than to buy a new one! Despite their cheapness, however, they are made of first quality HSS and I find they have an additional advantage in that they are shorter than the usual endmill and especially in the smaller sizes have less tendency to wander and so form an over-size slot. They are made from $\frac{1}{4}$ in. to $\frac{1}{64}$ in. in the long series, and $\frac{1}{16}$ to $\frac{1}{4} \times \frac{1}{32}$ in. in the short and ball-nosed types, with $\frac{1}{4}$ inch shanks, and 1 mm to 6 mm at similar intervals (6 mm shanks) in the metric range.

However, there is little point in having an endmill that will cut true to size if it is held in a three-jaw which runs out even only a thou or so, and it is worthwhile making a proper holder for them. You can, in fact, buy one but the cost of this may make a nonsense of the cheapness of the cutters! A No. 2 MT drill-chuck arbor is a lot cheaper. Fig. 4.12 shows the holder. The snag is that

such an arbor is all taper and not easy to hold while drilling for a draw bar. So you don't need a draw bar – you don't use one on your drilling machine? Don't you believe it! When drilling, the thrust will hold the taper into its socket, but even so it is not unknown when drilling brass to have the drill take charge and pull it out. Even more so when milling, for the one inescapable feature of the milling operation is the vibration set up in the process.

I have had many cases where small endmills have been slowly pulled out of plain spring collets, too, for the forces at the cutting edge of an endmill have a downward component in some cases, so a draw bar is essential.

The procedure* is to turn a parallel part on the arbor, both to hold it and to support it when drilling. Clean out the taper in the headstock, and that of the arbor, and first machine a parallel section on the taper that normally fits the drill chuck. I suggest about $\frac{3}{8}$ in. long – just enough to get a grip on it with the three-jaw. Fit the latter and hold this parallel collar in the chuck, supporting the free end of the arbor with the tailstock centre, and turn a short parallel section at the end of the taper, wide enough, this time, to suit your fixed steady. Get a good finish in both cases, and remove any sharp edges at the junction of taper and parallel.

Now set up your fixed steady and using the tailstock centre as a guide (or any other means to your taste) adjust it to hold the arbor true. Withdraw the tailstock, and then part off the tang. Refit the steady and face the end, centre deeply and drill tapping size to suit the preferred draw bar. Anything over $\frac{1}{4}$ in. will do, but I use $\frac{3}{8}$ in. Whitworth, which

*See also pp 52 and 75.

120

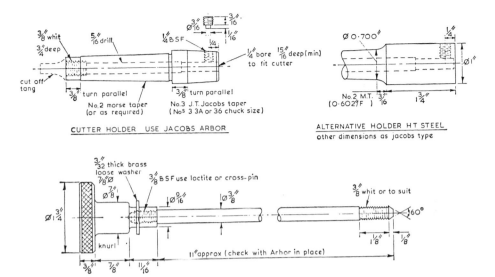

Fig. 4.12 Details of the milling cutter holder and draw bar. Dimensions are for $\frac{1}{4}$ in. shank Clarkson FC3 cutters.

is $\frac{5}{16}$ in. tapping drill. I say "drill deeply" – you really need a hole right through, but don't take the tapping drill in further than to within $\frac{1}{16}$ in. of the end in the chuck – the hole at that end must be bored to suit the cutter. Now tap the hole you have made, about $\frac{5}{8}$ in. deep or so. Make a draw bar if you have not got one – this is a simple job and lots of chaps just use a length of screwed rod with a nut and washer at the end. However, the sketch shows the one I use on the Myford Super-7. I will not give instructions as it is easy enough to make. It should screw into the arbor about $\frac{1}{2}$ in. Note the brass washer, which protects the end of the lathe mandrel.

The arbor can now be mounted in the headstock taper (clean first, please) and held by the draw bar. The centre already in the end should be true, but if you have doubts turn it out and make a new one. Drill $\frac{7}{32}$ in. very carefully; deep

enough to break into the existing hole, and then bore 1 in. deep until the cutter is a smooth sliding fit. Note that there will be a tolerance on the shanks of the cutters; mine all lie between less $\frac{3}{4}$ and less $1\frac{1}{4}$ thou of $\frac{1}{4}$ in. dia, so choose the largest. If you only have one cutter then I suggest you bore to a close slide fit to 0.250 in. dia. Now, if you have doubts about boring such a relatively deep hole in such a size, drill $\frac{7}{32}$ in. as before, right into the other hole. Bore as deep as you can until a letter D or 6.2 mm drill will just enter; put this drill right through, feeding gently, and follow with a $\frac{1}{4}$ in. reamer. Put the reamer through using the tailstock drill chuck, run the lathe at about 300 rpm, use lashings of cutting oil, and 'woodpecker-feed' the reamer, in and out about $\frac{1}{8}$ in. more each cut. This will give a true hole, but I would prefer to bore it, as you really need it *slightly* under-size to get a good fit to the shank of the cutter. Whichever way you have done it, lightly countersink the entrance to the hole.

Remove from the lathe and drill and tap for the setscrew (the cutters have a flat to suit). I strongly recommend that you turn down the end of this to the shape shown in the sketch; the cutters are, of course, hard, and unless you have a reasonable size of pip on the end the screw will soon burr over and jam. If you use an ordinary steel screw you may find it worth case-hardening the end. I should have mentioned that these cutters are also available in metric dimensions. If you decide to go for these, then the bored hole will be 6 mm instead of $\frac{1}{4}$ in. dia; everything else can be the same.

If you do not have a Jacobs arbor of the right size and cannot get one, then you may have to make the whole issue. In that case I suggest you enlarge the head as shown as an alternative. This gives a bit more projection from the mandrel nose and so reduces the overhang of the saddle over the gap in the bed. It adds to the overhang of the cutter, of course, but the head of the holder is pretty stiff. I have given the standard taper on the sketch, but it is better to set up a dial indicator and adjust the top-slide either to match the internal taper of the mandrel or that of an arbor. If you do the latter, make sure that the tail-stock is set to turn parallel – really parallel – before setting the arbor between centres.

Finally, if you use this holder – or any other device that fits the headstock mandrel taper – you ought to have a protective cap to go over the screw and register, just in case anything drops onto it and makes a burr. One made of boxwood or lignum vitae will do. It does not need to screw on, just make it a *tight* fit so that it does not want to go on, and make half-a-dozen sawcuts in it to give a bit of spring. On all precision lathes which are both bored for collets and screwed to receive chucks such a fitting is supplied as standard with the machine. I am a bit surprised that Myford do not market the thing as an optional extra!

A micrometer scribing block

I have, for more years than I care to remember, used a little scribing block or surface gauge made by *HJORTH* which had a number of advantages over the normal adjustable type, but it was rather small – the scribing point would not reach up to lathe centre-height except at a steep angle, for example. I had for some time intended to make a slightly larger one. The photo Fig. 4.13 and Fig. 4.14 show the general arrangement. It will be seen that the scriber is carried in a slotted 'slider', movement in this slot providing coarse adjustment. The slider is adjusted in turn by the knurled nut working on a 40 tpi screw, the nut being calibrated in 0.001 in. divisions. This provides just over $\frac{1}{4}$ in. movement. A further facility is provided in the form of a hole in the base; the bent scriber can pass through this, thus enabling marking out to be done on a vertical face below the instrument. Finally, the two sides of the base are made dead square to the front, and the blade dead square to the base, so that it can be used as a block square.

The dimensions are such that in mid travel of the micrometer the scriber is almost horizontal at $3\frac{1}{2}$ inches, Myford centre-height; it could, of course, be made taller, but this depends both on availability of material and the capacity of the milling equipment. The total

Fig. 4.13

range is from $7\frac{1}{2}$ in. above the base to $2\frac{1}{2}$ in. below with the long scriber. For those using a vertical slide the size shown is about the limit. The material is bright drawn mild steel. There is no point in using ground stock as, apart from the cost, this confers no advantage unless it can be hardened, and if hardened the construction would require a spot-grinding process, for which few model engineers have either the equipment or the skill. The use of BDMS does, however, need some extra machining to get over its well-known lack of flatness and tendency to distort when machined. Manufacture presents few difficulties but does need some care if the block square feature is to be retained. The decorative recessing of the blade, referred to later, is optional and would be difficult without a swivelling vertical

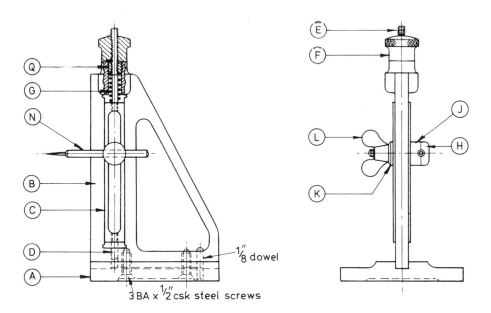

Fig. 4.14   Micrometer scribing block general arrangement.

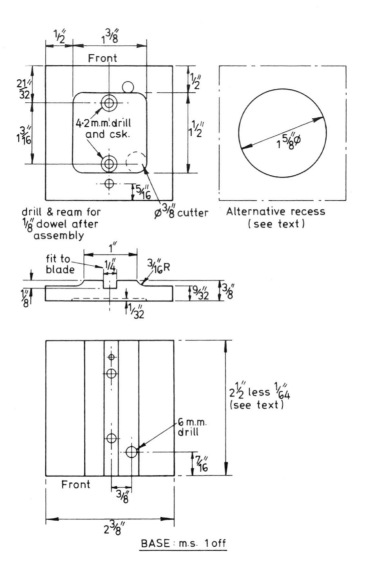

Fig. 4.15

slide or a rotary vice if a miller is available.

Preliminary work
The first step is to prepare the material for base and blade, see Fig. 4.15. This involves machining both faces of the stock, both to relieve stresses and to get a flat face. I should say here that my own instrument was not so treated on

the blade, and this I regret (it should have been stress relieved first). The distortion was minimal, but the surface finish was far from good enough. Taking the blade first, cut off a piece of $\frac{5}{16}$ in. stock, $2\frac{1}{2}$ in. wide by a shade over $3\frac{3}{4}$ in. long. Roughly square off the ends and then set in the four-jaw and take off about $\frac{1}{64}$ in. Reverse, set back on parallels and take off $\frac{1}{32}$ in.; then reverse again and, taking some care to see that the plate is hard back on the parallels, bring the thickness down to size. This should be a shade over $\frac{1}{4}$ in. rather than under. Take care during this operation – fit soft packing under the jaws to avoid marking the edges of the plate.

Carry out the same operation on the base, starting with material thicker than the $\frac{3}{8}$ in. called for; I had to use $\frac{1}{2}$ in. In this case there is no need to take two nibbles at the first side. Now, the drawing shows a rectangular recess on the underside, see Fig. 4.16, but you can make this circular if you like, and this can be done while in the four-jaw. The object of this recess is to reduce the amount of fitting needed when making it truly flat later. You now have the raw material for the job flat, and with reasonable luck, the pieces will remain so! Draw-file the edges, keeping them square to the face, and square off the ends properly, making the $2\frac{1}{2}$ in. length of the base about $\frac{1}{64}$ in. less than the corresponding dimension on the blade (Fig. 4.17). Measure the thickness of the blade and if one end is more parallel than the other, mark this as the bottom edge. Check the squareness of the two ends of the base with one side, and mark the most square as the front. Similarly, check and mark the blade, selecting the side which is most square to the base as the front.

Base (Fig. 4.15)
Set the vertical slide facing the chuck

Fig. 4.16 Underside of the base showing the recess.

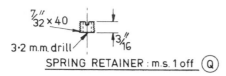

$\frac{7}{32}'' \times 40$

3·2 m.m. drill

$\frac{3}{16}''$

SPRING RETAINER : m.s. 1 off Ⓠ

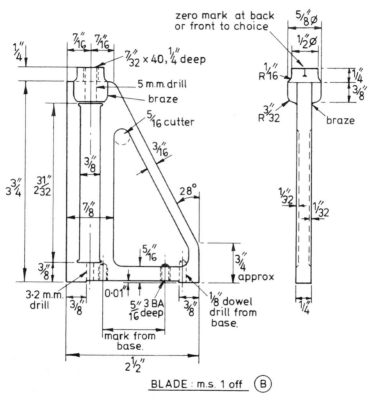

zero mark at back
or front to choice

$\frac{5}{8}\emptyset$

$\frac{1}{2}\emptyset$

$\frac{1}{4}$

$\frac{7}{16}''$ $\frac{7}{16}''$

$\frac{7}{32}'' \times 40, \frac{1}{4}$ deep

5 m.m. drill

braze

$\frac{5}{16}$ cutter

R $\frac{1}{16}''$

R $\frac{3}{32}''$

braze

$\frac{3}{8}''$

$\frac{3}{16}''$

$\frac{1}{4}$

$\frac{3}{8}''$

3$\frac{3}{4}''$ 2$\frac{31}{32}''$

28°

$\frac{7}{8}''$

$\frac{1}{32}''$

$\frac{1}{32}''$

$\frac{5}{16}''$

$\frac{3}{4}''$
approx

$\frac{3}{8}''$

3·2 m.m. drill

$\frac{3}{8}''$ 0·01'' $\frac{5}{16}''$ 3 BA $\frac{3}{8}''$

$\frac{1}{8}$ dowel drill from base.

deep

mark from base.

$\frac{1}{4}$

$2\frac{1}{2}''$

BLADE : m.s. 1 off Ⓑ

Fig. 4.17

and take care to get it dead square across the lathe bed. Mark out for the slot, and two lines $\frac{3}{8}$in. each side of it. If your vice will take the piece, grip in this with about $\frac{5}{32}$ in. projecting; if not, you will have to use clamps and do the job in two bites. Set the clamps on the longer sides, and set the job with the scribed lines horizontal. Machine the slot, using a $\frac{1}{4}$in. slot drill if your chuck is

dead true; otherwise you must take two bites with a $\frac{3}{16}$in. or $\frac{7}{32}$in. tool. The blade should be a tight, but not drive, fit. Take a depth of cut each time not more than $\frac{1}{16}$ in. and keep the chips cleared.

Change to a ball-ended slot drill, say $\frac{3}{8}$ in. dia (the size isn't critical) and machine along the two scribed lines at the sides of the slots. Get a good finish here. If using the vice you can then mill the flat surface to blend in with this radius, but if using clamps you must fit two more to hold the job and then remove the existing ones before milling the face. Finally, take a very light skim – only a few thou – off the long sides. There is no need to go the full thickness of the plate, as the object is to get a witness to file to later. If you are using a vice you may be able to do likewise on the front and back faces too, to serve the same purpose.

Remove from the machine and take off all burrs. With a small fine-cut file and some care go round these edges and file truly square both to the front face and to the underside. Clean out all toolmarks on the top surface, and draw-file the flat faces each side of the slot. Leave the underside as it is for the present – this must be bedded to a surface plate or piece of plate glass later. Mark out for and drill the holes, but leave that for the $\frac{1}{8}$ in. dowel until after assembly.

Blade (Fig. 4.17)
Mark out for the slot and the outline, but *do not cut the outline to shape just yet*. Set on the vertical slide with the long side horizontal, using a dti to ensure that the 'front' is true across the bed. You will need packing behind as the slot drill must go right through. I strongly recommend that you take two bites at the slot, machining each side of it separately, as you need a good finish here. Make the slot a shade less than the nominal $\frac{3}{8}$ in. just enough to allow you to clean up the finish afterwards with a fine file. The rounded ends will be squared up later.

Now, if you are using a swivelling slide you can tackle the recess. It need be no more than $\frac{1}{32}$ in. deep. Using the cross slide and the vertical slide feedscrews go round the 'square' edges with

a $\frac{5}{16}$ in. or $\frac{1}{4}$ in. slot drill first, just up to the marks, and then chew out almost to the sloping back line. Set over the swivel on the slide so that the marked-out line is either vertical or horizontal – the latter is more convenient – and run the slot drill along this line, taking care to get a good blend at the ends. The next step is a refinement – you can do it or not as you wish. After machining the sloping side of the recess I took a cut about $\frac{1}{16}$ in. deep along the *outline*: this gives a guide for later sawing and filing. Further, at each end of this cut I plunged the slot mill right through (taking care that the packing didn't foul) to aid in the setting up which follows.

Reverse the piece; if you have made the two holes just mentioned you can align by offering the slot drill to these holes, otherwise you must use care and normal marking-out procedures. Repeat the operation for this side, but do the sloping side first.

You can now saw off the unwanted metal and finish the back to suit your taste. Draw-filing is best but don't spend a lot of time at present, as a brazing operation is to follow soon. This is to unite the top boss to the blade, so make this boss next. A straight turning job, but it *is* important that the slot be dead central. The trick here is to drill No. 10 while in the chuck – don't tap the hole – and finish all else. Part off and remove burrs. Set up in the machine vice on the vertical slide and align until a No. 10 drill held in the chuck enters freely. Replace with a 5 mm drill and open out the hole, then fit your $\frac{1}{4}$ in. slot drill and machine the slot. It then can't be other than properly aligned. Take care to tap the $\frac{7}{32}$ in. $\times$ 40 tpi hole squarely.

Set the boss on the blade and check that the hole is central to the slot. Flux well, and braze together with *Easyflo*

No. 2 or to your choice. Don't use too much alloy – just enough to make a nice fillet. You can now clean up the whole of the blade, and file a little relief along the bottom slide of the triangle to within about $\frac{3}{8}$ in. of each end, and about 0.010 in. deep. Before setting aside, check the squareness of the front and bottom edges and if they are not almost exactly right, save time now by filing the *bottom* side true. Check the fit of the blade to the base, and if necessary very gently ease the slot in the latter. The two parts should fit closely but not be jam tight. A tap with a (small) plastic mallet should suffice. Pop through the two csk holes in the base and then drill the blade and tap. I have called for 3 BA but 4 BA or 2 BA will do. Take great care that these holes are square to the blade for if they are 'slantendicular' they will pull the blade over. Don't assemble permanently just yet, but don't forget to pop through and drill the hole 3.2 mm.

Slide (Fig. 4.18)

This again is best filed or milled down from larger section stock, but make the $\frac{3}{8}$ in. slightly over-size by just a few thou. Set up in the vertical slide, and here a precaution is needed. Even if your vice is long enough to accept the full length there is a risk of the slot closing in, so grip it between two longer parallels with a piece of cigarette paper between each parallel and the workpiece, just at each end. Mill out the slot with a $\frac{7}{32}$ in. slot drill, taking light cuts to get a good finish. Remove all burrs and refine the finish in the slot; the width of this is not critical.

Now fit this to the blade. Square off the ends of the slot and then file both or either of the slot and slider until the latter is a smooth fit. It must travel freely, but without any shake, and

should have $\frac{5}{16}$ in. of travel. Recess the ends of the slot as shown on the drawing. Now spot through the 5 mm hole at the top and the 3.2 mm at the bottom of the slot, to mark the slider. Taking care to ensure that the holes are true, drill and tap these $\frac{1}{8}$ in. × 40 in. tpi and then lightly countersink the holes.

I suggest you now make the lower peg and the adjusting screw (Fig. 4.18) and, with the slider in place, offer these up and make sure that they don't interfere with the smooth travel. If they do, enlarge the 3.2 mm hole at the bottom a bit at a time until all is free. We may have problems at the top later when the spring is in place, but we will deal with that when it arises.

Adjusting knob (Fig. 4.18)

This is a straight turning job, but it *is* important to see that the tapped hole is true. Hold the shank of the tap lightly in the tailstock drill chuck and take it slowly, using plenty of cutting oil. Now for the engraving. Purists will set up their dividing heads and do it that way, but this is quite unnecessary if you have a graduated handwheel on your leadscrew (there is, after all, no point in using super-accurate methods of making the index when the screw it controls is a commercial 40 tpi thread). Set a sharp-pointed screwcutting tool first to make the ring round, then lay it on its side at exact centre-height to form the graduations. Set up for 8 tpi, when the leadscrew and mandrel ratio is 1:1. Apply friction to the handwheel and rotate the chuck until the index reads zero. Make a long mark, about 0.010 in. deep. Going the same way round, again rotate the chuck with friction applied to the handwheel until its index reads '5'; make a short mark. Rotate again to '10' and make another; and so on. If you

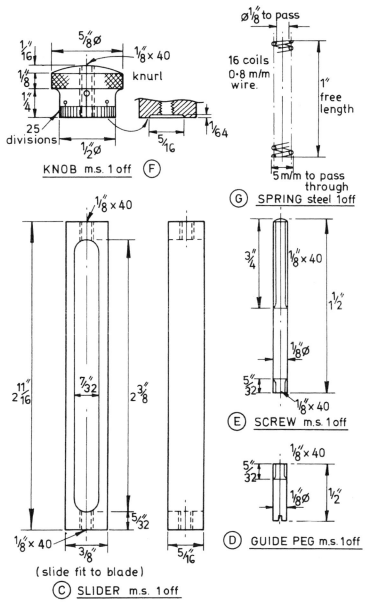

ø$\frac{1}{8}$" to pass

16 coils
0·8 m/m
wire.

1"
free
length

5 m/m to pass
through

(G) SPRING steel 1off

$\frac{1}{16}$"

$\frac{1}{8}$"

$\frac{1}{4}$"

$\frac{5}{8}$ø

$\frac{1}{8}$" × 40

knurl

25
divisions

$\frac{1}{2}$ø

$\frac{1}{64}$

$\frac{5}{16}$

KNOB m.s. 1off (F)

$\frac{1}{8}$" × 40

$\frac{3}{4}$"

$\frac{1}{8}$" × 40

$\frac{1}{2}$"
1

$\frac{1}{8}$"ø

$\frac{5}{32}$

$\frac{1}{8}$" × 40

(E) SCREW m.s. 1off

$\frac{1}{8}$" × 40

$\frac{5}{32}$

$\frac{1}{8}$"ø

$\frac{1}{2}$"

(D) GUIDE PEG m.s. 1off

$\frac{11}{16}$"
2

$\frac{7}{32}$"

$\frac{3}{8}$"
2

$\frac{5}{32}$"

$\frac{1}{8}$" × 40

$\frac{3}{8}$"

(slide fit to blade)

(C) SLIDER m.s. 1off

$\frac{5}{16}$"

Fig. 4.18

overshoot, you must go well back and approach the handwheel index mark always from the same direction. Make a slightly longer mark at the 5th division (25 on the handwheel) and longer still at the tenth (50 on the handwheel) and so on. Remove the burrs, polish, and this will fill the engraving with dirt and make it easy to read (you can use paint if you like, but dirt is cheaper).

Spring and retainer
The retainer (Fig. 4.17) should present no problem, but you will have to ensure that the slot shown will fit whichever of your screwdrivers will enter a $\frac{7}{32}$ in. × 40 tpi hole! The spring (Fig. 4.18) is another matter. It must be fairly strong, and yet be a close fit (but not tight) on the $\frac{1}{8}$ in. dia spindle. In addition, it must clear the 5 mm hole in the blade. This is a tight specification, and I solved it by seeking out a spring which would fit from my stores. However, I did check whether it could be done and found the following to be necessary. The 'mandrel' was of hardened and tempered (light blue) silver steel, turned down to 0.105 in. dia first. The spring wire (0.8 mm dia) was brought to dull red and cooled slowly. The mandrel had to be supported by a female centre, and the spring wound close coiled. It was then pulled out (after cutting off 16 coils and grinding the ends flat) to 1 in. free length. After heating to red and quenching in oil it was tempered by heating inside a copper tube until the quenching oil flashed off, keeping the tube rotating all the time. The spring was not quite as stiff as the one actually used, but was quite adequate. However, there is an alternative and that is to alter the hole in the boss and the thread to suit the spring; there is no reason why the thread should not be $\frac{1}{4}$ in. × 40 tpi, and

the hole in the blade can go up to 5.5 mm if you take care.

You must now offer up the slider in the blade with both the locating peg and the spring, adjusting screw and screwed plug all in place and make sure that nothing binds. If it does, then you must adjust with a fine rat-tail file.

Scriber clamp (Fig. 4.19)
The only problem here is to get the hole for the scriber in the centre, and the only methods are (a) care and (b) make and use a cross-drilling jig. I used the former as being quicker and more of interest. The collar and the boss were gripped in the drill vice and a start made with a small drill at the junction. This start was enlarged bit by bit, each time adjusting the position so that the drill aimed true – you will find it tends to wander sideways. The $\frac{1}{8}$ in. drill was then put through and the collar will almost, but not quite, fit both ways round. The face of the collar is then filed down so that the hole will grip the scriber when all is tight.

The nut in my case is a commercial wingnut cleaned up; preferable to a knurled nut as it can be screwed up tighter.

Scribers (Fig. 4.19)
The drawing shows two; one is simply a piece of $\frac{1}{8}$ in. silver steel with an old-fashioned gramophone needle fixed in the end with *Loctite*. The end should, of course, be made dead sharp on an oilstone, as such needles are formed to a very small radius to suit the old 78 rpm records. I use this for all precise work, as it is very sharp and stiff (a sewing needle might do, but I haven't tried that one). The other is a 'bent' one. The silver steel is first rough turned to shape at the ends, then bent (cold) and

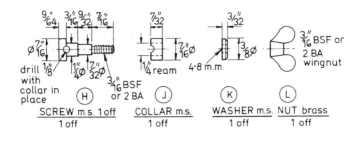

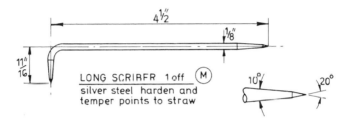

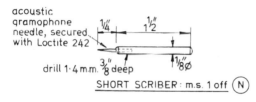

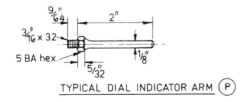

Fig. 4.19

hardened and tempered to straw. The points were then formed on the fine wheel of the grinder and finished off with an oilstone. The third attachment is to carry a dial indicator. The one shown is to suit my Last Word indicator, and you would have to alter yours to suit your instrument.

Assembly and adjustment
The base should be made dead flat by scraping to a surface plate (or sheet of plate glass) with marking blue. If it is very much out you may have to start proceedings with a file. I made mine flat to a dark blue (that is, with thick marking blue) before assembly, and

131

finished to light blue afterwards. Assemble the blade to the base, screws tight, but without the slider etc. Check that the blade front stands square to the base. It should project just a trifle at the front, and if you are not going to use it as a block square, this is well enough. The back end will also project; this should be filed flush with the base and finished off nicely. Having checked that you can see no daylight between your square blade and the blade of the instrument, check that it is straight with marking blue on the surface plate and correct the scraper. As you do this, keep referring also to the square, in case you take it out of true. Once satisfied, that is that so far as use as a surface gauge is concerned. Drill and fit the dowel peg, and apply paint to your taste.

However, if you want to use the instrument as a block square you must go further. Apply the tests as above, but don't scrape the front blade – instead, very carefully file it until it is flush with the base. Every few strokes with the file check the squareness and it will pay also to check on the surface plate, too. Once you find blue in traces right along the blade *and* along the front of the base, revert to the scraper and bed in until you get about 70% marking. *Check with the square all the time.*

Now, from the front face, make one side square to this, again filing at first and then scraping. (By now your fingers will be fully impregnated with marking blue – one of the prices of accuracy!) Once this is true, repeat the operation on the second side of the base. Thereafter proceed as before with paint as desired.

You now have an instrument which will not only serve as a scribing block, and which will avoid the need for setting up the dti for many applications, but which can also be used to check squareness in three dimensions (on stretchers in locomotive frames, for example). It will not be to toolroom accuracy, of course, but it is a great deal better than trying to hold two squares to a job at the same time – with no hands free to do anything else! For normal marking out the device has the great merit of stiffness, and the pointer doesn't move back and forth when adjusted up and down. Within its limit it serves as a vernier height gauge when marking out from a centreline (like the slot in the base, for example) and though a 40 tpi die can't be expected to make a precision leadscrew the graduations are sufficiently accurate for their purpose. The bent end of the scriber can be used to check the height of work by feel (which can be surprisingly sensitive) and any error can be measured (Fig. 4.20). The facility for marking out 'downstairs' with the scriber passed through the hole in the base may not be needed very often, but is invaluable when the situation does arise.

In conclusion, I must make two things clear. First, I make no claim for originality of design; that was due to the company called *HJORTH* many years ago. But I have, I think, made a few improvements, quite apart from the increased height. The original had a very limited micrometer adjustment, and the knob was so small that one-hand use was difficult. Second, I have in this case departed a little from my principle of describing only processes I have done myself. My own instrument was made using a vertical milling machine with a graduated rotatable machine vice. However, I *have* checked that the work can be done on a swivelling vertical slide on the lathe provided

Fig. 4.20 Checking the height by feel (see text).

Fig. 4.21 Using the scriber point on a surface below the base of the instrument.

Fig. 4.22   The short scriber fitted.

that sufficient care is taken in setting up and that depths of cut are kept down in proportion to the lack of stiffness in the set-up. Those who have milling machines can, of course, translate the operations I have described to their machine.

Dividing from the chuck

The centre-height device shown on page 58 or that just described can be used very simply for spacing out simple equal division – as may be needed, for example, for cylinder cover bolt-holes (see Fig. 4.23). A small spirit level is held on each chuck jaw in turn and a line scribed right across. This photo shows a four-jaw self-centring chuck, giving 4 holes. With a three-jaw (Fig. 4.24) divisions can be made for 3, 6 or

12 holes – the latter by setting the level first on jaws at the front and then on jaws at the back of the lathe.

The level seen has a slot in the base, so that it can be set on an angle-square if necessary e.g. 45° in conjunction with a four-jaw chuck to provide 8 divisions.

Straightening copper tube

The copper tube division of the 'Tubal Cain stores' – the top of a 6ft steel shelving unit – had got into a mess, with all sorts and sizes mixed up. True, the copper was all dabbed with yellow paint and the brass with red, but it *was* a mess and a couple of hours was spent in sorting it out, at the end of which time there was left on the bench a mixture of bent and contorted tube which had been used for all sorts of purposes over the years. But for the price of copper these days, fit only for scrap. So, to straighten it.

First, the nipples were removed, and any solder coated ends sawn off – the slightest trace of this would be fatal to any subsequent brazed joint. A few of the less badly bent pieces were straightened with the fingers and put to store. Of the rest, quite a lot was straight but not bent if you follow me – it had a few bends in it but no kinks and the bends were regular. These were first annealed by heating to red. No need to quench – this does no good other than saving time on copper. After getting as straight as possible by hand the device seen in Fig. 4.25 was used. This is a 'gag-press' though in this case the press part is simply my $4\frac{1}{2}$in. bench vice. The grooves are made to fit various sizes of tube – the drawing shows only three and, in fact, it is best (and easier to make) if those for small tubes, which need not be so long, are on one piece of wood

Fig. 4.23 Setting out holes in a pitch circle using a level on the chuck jaws.

Fig. 4.24 As Fig. 4.23 but in the three-jaw chuck.

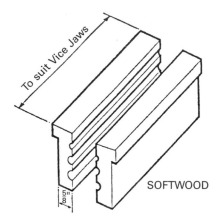

Fig. 4.25   Gag jaws for use when straightening copper tube in the vice.

and the others on another. The holes are drilled in the solid piece of wood, any sort will do, and then split with a tenon saw.

This will deal with most problems but I should imagine that most of you will find that you are left with a couple of revolting pieces of knitting so bent that they look hopeless. Not so. Had they been copper wire I should just have got hold of one end in the vice and the other with a mole wrench and given a good heave. This isn't possible with any but the thinnest copper, as I haven't enough pull! So, try this way. If there is a nipple brazed on one end, leave it there; otherwise, braze or solder on a block with a hole in it. Slip a washer over the tube and grip the other end in the three-jaw of the lathe. Pass the tube through the aperture of the toolpost with the washer against the tailstock side, engage the half-nut and rotate the leadscrew handwheel to pull on the tube. You may find it necessary to resoften after the first pull, but even with quite large diameters – up to $\frac{5}{16}$ in.

o/d – it works wonders. With pieces that are too long it may be necessary to straighten as best you can by hand and thread it through the hollow mandrel or even cut it in two. But with prices the way they are, two straight bits 15 inches long are better than a two-foot corkscrew.

Cutting and threading copper tube
The cutting forces operating on the tool point are now fully appreciated by most model engineers, and many take full advantage of modern techniques as applied to tool shape and support in the lathe. What is not so fully understood is the fact that these forces act on the workpiece as well – 'if the horse pulls the cart, the cart pulls the horse', a graphic restatement of the law that action and reaction are equal and opposite!

These forces are seldom of much consequence, but when working on copper tube the radial and tangential components of the cutting force are usually sufficient to distort the material. This is especially the case when cutting a thread on thin-walled tube. Indeed, close observation of the behaviour of the material under the influence of a three-lobed die gives a very clear picture of the nature and direction of the forces concerned. Under the influence of a single-point tool in the lathe the effects are often catastrophic.

To overcome these difficulties, use may be made of a supporting mandrel. Fig. 4.26 shows such a device applied to screwcutting with a die. The diameter D is made to be a good, but not tight, fit in the bore of the tube, while C is a sliding fit in the die (in this connection, do not overlook the fact that a split die is adjustable). The length A should be equal to the length intended to be

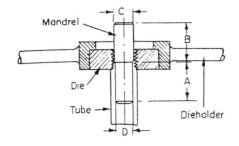

Fig. 4.26 *Support mandrel used in threading thinwall tube.*

screwed plus about 1.5D. B is immaterial, but is conveniently about 2C. Even with the thinnest material this device will prevent collapse of the tube, and has the added advantage that the die is partially guided to give a reasonably true thread. Removal of the mandrel is facilitated by the use of castor oil or grease on the plug.

Similar devices will be found useful in the lathe. A mandrel ought always to be inserted in the length of tube gripped in the chuck; well oiled and preferably with a screw thread on the back end to enable it to be started when withdrawing from the tube. Parting off is quite straightforward if the mandrel extends under the tool, and the use of the tailstock die-holder will present no difficulties. For other operations, the

arrangement shown in Fig. 4.27 will be found helpful. This has been used mainly for screwcutting operations, but is also invaluable when turning up collars brazed onto the pipe end when the tube is usually very soft. The pressure exerted by the back centre should be as light as possible, and the mandrel is long enough to take the pressure from the chuck jaws. In one particularly difficult job – turning down the end of a $\frac{5}{16}$in. $\times$ 30g tube to 0.005 in. thick – the mandrel was made a very good fit in the bore, and subsequently removed by heating. This led to the experiment of fitting a tube to a mandrel by shrinking so that it could be turned between centres. This is effective on tubes of reasonably short lengths, but leads to difficulties if the length is more than about 8 diameters.

The material of all mandrels should be steel and should be highly polished where it is in contact with the tube. My standard die-screwing mandrels are made of silver steel, hardened, and tempered to 300°C – dark blue. They are marked on the end with the tube outside diameter and thickness.

Awkward nuts
Nuts which are quite accessible on the full-sized prototype often present acute difficulties on a model simply because

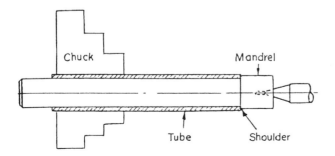

Fig. 4.27 *A mandrel to support tube in the lathe.*

we have not got scale fingers! And at the scale of my last model my 10 BA tube spanner is 4 feet long by 5 inches diameter. Try it this way. Put a very slight lead on the first thread of the studs and run the die over. Very slightly countersink the nut – just show it to a 60° centre-bit and put the tap through to take off any burr. Offer the nut to the stud with tweezers and then set the point of a scriber through the nut on to top of the stud. You can then 'persuade' the nut to make a turn or so with the points of the tweezers, after which it will stay put while you tighten up properly with a spanner.

Unfortunately many BA hexagons are now metric, and may be a slack fit in the end of the box spanner, so that when you try to pick up a nut and carry it in the spanner to the stud (assuming that you can get the spanner into the space) the nut falls out. The usual remedy suggested is to magnetise the spanner, but quite apart from the fact that some nuts are brass I find that the magnetism often draws the nut so far up the cavity that it won't reach the stud. A spot of plasticine (or even grease) will often do the trick.

Catching rings and washers
Some time ago I had a dozen or so little washers to make, about $\frac{3}{16}$ in. o/d and $\frac{1}{8}$ in. thick. At the beginning I lost about one in three in the swarf tray. Bring up the tailstock drill chuck, set a piece of stiff wire in it and feed it through the already drilled stock. The washers will just walk along the wire to be collected quite safely.

Chatter on boring bars
Occasionally one experiences chronic chatter when boring. It is usually due to an unfortunate coincidence between the natural frequency of the bar, the grain size of the metal in the casting and factors like depth of cut and cutting speed. (There can be other causes, like the shape of the cutting tool point, too.) It is not always convenient to alter cut or speed, but the natural frequency of vibration of the boring bar can sometimes be shifted enough to cure the chatter if a lump of plasticine or even *Blutack* is plastered on the bar towards the end. Lead strip can be embedded to add mass. It must be clear of the cutting point, if course, and it will be necessary to clear the chips from it fairly often, but it is quite effective. See Fig. 4.28.

My blackboard
This seems so simple, silly almost, but it may well be the most important aid to production (and accuracy) in the shop. Not very large, just a piece of plywood about 4 inch × 16 inch, fixed to the shelf above the lathe, the chalk living just above, *on* the shelf. On it I can note down the last cut when I am called

Fig. 4.28 Plasticine damper applied to a boring bar (see also Fig. 2.26, p 40).

away from the machine, or a rough sketch of the workpiece, a reminder that the tailstock is set over and a host of other things which must not be forgotten. Many visitors have remarked "Oh – I use a piece of paper" and, within days, ring up to say that they have made a similar board and found an immediate benefit.

It seems easy to make, but like all easy jobs there are tricks of the trade. Special blackboard paint can be obtained, but it's hardly worth it for so small a job. The secret is in the preparation. Damp the plywood and rub down well, after it has dried, with medium glass or garnet paper, to get a smooth surface. It won't stay smooth, but the initial treatment saves work later. Apply a priming coat and, when well dry, rub down and then re-prime. Rub down again and apply whatever undercoat you have – any paint let down with a little white spirit will do, but put on several coats and rub down well between each and after the final coat. Obviously if you have a matt black paint available this is easier than gloss for the final coats, but as you have to rub down anyway it isn't all that important. You will need several coats, and the final one should be rubbed down gently with wet-and-dry finishing paper, used wet, about grade 360 or 400. This should leave a satin-smooth surface which will take chalk well.

It may seem a lot of work for so silly a thing, but even if you use the proper blackboard paint you will have to go through the same procedure if the chalk isn't to slip on the surface instead of writing. My blackboard, in use for 25 years (Fig. 4.29) has never needed recoating – just an occasional wipe over with a damp rag. If you live in Wales you already have the ideal blackboard material – slate. Ordinary roofing slate will do and only needs smoothing with wet-and-dry using progressively finer grades (used wet) to bring it up to a superb chalking surface which will last a lifetime. Drill the holes with a rawldrill and fix it with screws, although a real slater would give just four taps with his piercing hammer!

Tailpiece
No – not an afterthought because I did not think of it at all! Just a device for meeting that ever-present job of centring

Fig. 4.29   My blackboard!

work to a 'pop' when using the four-jaw chuck – and so *very* simple. When asked for a photograph to adorn the front cover of this book, Mike Chrisp (technical editor of *Model Engineer*) set up a device of his own. One look at it and I went into shock reflecting on all the time I had wasted over the last 60 years!

The author's favourite device – 'a little piece of wood'!